# Yuwayol

Secret Phil

That you exist,
Against the odds.
That anything exists,
At all.
Two curious coincidences.
Too bizarre.

Yuwayol by Secret Phil

© 2018, Secret Phil

Self-published

ISBN    978-0-244-98998-9

# Contents

# Preface

A few years ago I came to the sudden realization that I was never going to get an answer to any of the questions which have bugged me throughout my life. They are a set of very simple and, I would have thought, perfectly reasonable questions which any thinking living creature ought to be asking. These questions go along the lines of – "What am I doing here?", "What is my purpose in life?" and "Where and what exactly am I in relation to everything anyway?"

We are all born, without having been consulted, into a universe we can barely understand where we are basically left to get on with it. But, get on with what exactly? With no obvious answer to this question we tend to just make something up and try and muddle through. As a result, understandably in the circumstances, along the way we often doubt ourselves and wonder if we are treading the right path in life. And through it all is the growing unease caused by the knowledge that we are all just going to die anyway.

Of course there are folk out there who don't seem to doubt themselves at all, folk who are very happy to tell us what they claim to know everything is all about, and are quite happy to tell the rest of us what we should be doing with our lives. But such people are not to be trusted. They have no authority in these matters, no more than you or I, whatever grandiose claims they make.

The difficulty is that there is no authoritative manual available to us to answer our reasonable questions. So, if you are like me, you go in search of the answers. You spend your life reading books and magazines, attending talks, watching and listening to programs, particularly (but by no means confined to) those dealing with weighty subjects such as cosmology, physics, neuro-biology, phycology, religion, evolution, philosophy death and love. You plunge deep into your own psych, trying to find out what makes you tick, contemplating, meditating and even praying, all the time being wary lest your own biases and assumptions cloud your judgement as you pursue the truth. You observe those around you, those you love, friends, family, but also strangers, watching and listening for those pearls of wisdom which can only come from human interaction. You also seek solitude as you attempt to put all your thoughts together, taking long walks perhaps, or gazing at the night sky before bed time, or soaking for hours in a once hot bath when the water has long cooled.

All of these things I have done, and much I have learned because of it. I believe that I am a more considered, calm and balanced human being that I would otherwise have been. But it has been a lonely pursuit. I believe that even those close to me have little idea how much time I have spent devoted to my pursuit of answers to my simple, reasonable questions. I keep most of it to myself, sharing my findings and thoughts only sparingly. I'm not sure why this is, for some reason I just find it easier this way. I am a secret philosopher.

And then a few years ago, as I have said, I arrived at the conclusion most of my wiser fellow humans came to much earlier than I did (probably saving them a lot of bath water!). That is, that I could never know the answer to my questions. But it was the reasoning by which I arrived at this conclusion that was sobering. It wasn't just that I'd reasoned that none of the answers I'd contemplated ultimately stood up, nor was it so much that our understanding of the universe, or ourselves, as impressive as it is, just isn't sufficiently advanced to provide us with the answers. No, the central problem is that I could see no solution, *hypothetical or otherwise*, which could answer my questions to my satisfaction. And the reason for that is that, whatever general approach you took, paradoxes always introduced themselves into the arguments.

For a time this conclusion depressed and frustrated me, but then, like everyone else, I learnt to accept that I would never get to know the unknowable.

It is a curious thing how so often in life, when we finally face up to something we have feared to contemplate, the fear turns out to be unfounded and a new door opens for us. Actually, I don't know if my acceptance somehow freed up my mind, perhaps, and led to the thinking that I eventually began to refer to as Yuwayol, but that is the way it felt.

And it began with me idly contemplating one day (I still take the occasional bath) that, if there was a God, and if God really was the originator of all things, the God who it was claimed had described himself to Moses thus – "I am who I am", then wasn't it reasonable to suppose that, starting from scratch and just finding oneself being there and simply existing for no particular reason, God, however mighty he may be, would surely have exactly the same set of questions about his existence that I have about mine, with presumably no more resource to answer those questions than I have.

And that was followed by the following thought: Well, maybe this scenario is one that is actually being played out, except that maybe it isn't God, maybe…

What followed was a paradigm shift in the way I viewed all that I know, and Yuwayol (though not at that stage named as such) was born. And as initially outraged as I was by the idea, I was startled to realize that, if true, the paradoxes thrown up by every other line of reasoning had been resolved.

Surely, I thought, someone somewhere must have considered this before and written it down. I scoured the internet searching for phrases such as "we are all one" or "we share the same soul", but in spite of trawling through many essays, forums and blogs I drew a complete blank.

There are certainly a number of stumbling blocks on the way to understanding Yuwayol. Primarily, it involves understanding that our primate based view of the world could well be misleading us in many ways. It also requires that we appreciate why questions such as "Who am I?" and "Why does anything exist?" have caused so much head scratching throughout human history. But these issues can be dealt with reasonably easily, and I decided to write down my thoughts, even though I had no clear idea what I would do with the result. I just felt this was far too important an idea to just keep locked away in the confines of my skull.

Whilst my secretive impulses delayed me in this endeavor I nevertheless finally commissioned the website, yuwayol.com, where my thoughts could be browsed by anyone who has means to access to the internet. In doing so I adopted the *nom de plume* Secret Phil for the first time.

However there are many who prefer the convenience and feel of a book, whether it be in paper or tablet form. Hence this publication, which is largely based on the contents of the website, edited slightly as appropriate to handle the new format, and includes this preface.

Whilst we cannot be certain of anything, for me these days Yuwayol doesn't just make sense, it strikes me as intuitive and, at times, almost obvious. So much so that I have to keep pinching myself when reflecting that its conclusions have not already been arrived at previously. It is my hope that, after reading this book, even if you are not convinced, you too will agree that Yuwayol is at least a reasonable philosophy. But if not, I would like to thank you for reading and hope, despite this, you will still be able to find something from the various topics covered in these pages that is of interest to you.

*Secret Phil (September 2018)*

# Chapter 1

## Introducing Yuwayol

**What is Yuwayol?**

Yuwayol is the philosophy which claims that we are all the same thing. Your self is the same as my self. As we view the world it is true that we do so from different viewpoints, as individuals defined by our experiences, feelings and desires. Nevertheless, when you look into the eyes of another, it is in fact your self that is staring back at you.

Yuwayol bases this assertion on the idea that the way we see the world is, in certain regards, misleading. In particular the world we live in should not be regarded as a great big space, full of things, moving through time. Rather, it should be regarded as single entity and each of the things we observe within it,

*...when you look into the eyes of another, it is in fact your self that is staring back at you.*

including ourselves, are merely different aspects of this entity.

The word 'Yuwayol' is an amalgamation of the English words 'you' (phonetically spelt 'yuw'), I ('ay') and 'all' ('ol').

To be clear, when I say we are all the same thing, this is not meant in the loose way that we are all similar or that we share a common cause or experience. Indeed, our experiences often differ greatly and are very unique to us as individuals. But Yuwayol means that we are *actually* all the same thing.

It means that in a real sense I am you and you are me. The entity that you consider to be yourself, what you might think of as the inner, deepest you, reading these words, is the same entity that is me, writing them.

Whilst you read this you have a name that you, and others, refer to you by; you have feelings, memories, thoughts and all the things you hold dear, and not so dear. But when you are me, writing these pages, you have my name (though for the purposes of this book I will be signing myself off as Secret Phil), my feelings and memories and thoughts. Of course this is not just the case for we two, you are everyone else also, as I am, and everyone else is us. By everyone I mean all living beings who have ever lived or ever will. And if there are living beings on the other side of the universe, I mean them too.

Therefore, Yuwayol concludes that whenever you are kind or helpful to someone other than yourself, you are literally being kind or helpful to your own self. Similarly, by doing anything to cause pain or suffering to someone other than yourself, you are directly causing your own self to feel that pain and endure that suffering.

*This is an extraordinary claim, I know, and quite likely sounds outrageous.*

This is an extraordinary claim, I know, and quite likely sounds outrageous. And, simple though the idea might be, it takes a while to get your head around it, perhaps a lifetime, but it can potentially revolutionize the way you perceive both yourself and your place in the universe. The purpose of this book is to make the case for Yuwayol, to explain why the idea is at least worth your consideration, and to discuss what I think it means for all of us. This is done in the forthcoming chapters (including this brief introduction) which follow on from here.

**Two Key Questions**

As we shall see, driving the reasoning of Yuwayol are two key questions. They are not new. People have been asking them for as long as there have been people around with enough spare time to think about such things. These questions can be expressed in various ways. I have chosen to express them as follows:

- Why does anything exist?
- What am I?

But be warned, if you have never concerned yourself too much with these questions (and in many cases even if you have) you may find there is more depth to them than you think.

I am reminded of Douglas Adams' marvelous tale, *A Hitchhiker's Guide to the Galaxy*, in which a race of hyper-intelligent pan-dimensional beings build a giant computer, named Deep Thought, to answer the question of 'life, the universe and everything'. The answer Deep Thought subsequently reaches (as you will know if you are familiar with the story, in case you haven't I won't say too much), hilariously creates great consternation for the computer's makers. But the problem, as Deep Thought itself points out, is that nobody really understands the question.

We have a similar problem: in order address these topics, we need to understand exactly what it is that we are asking. If you don't get the meaning of our simple looking key questions, it is likely you won't understand why Yuwayol makes any sense.

The next two chapters are devoted to appreciating these two questions and I hope will give you an appreciation for their profundity and just why they have taxed philosophers for many centuries.

### Yuwayol is Not Mystical

You could be forgiven if you are wondering if I am touting some form of supernatural or mystical explanation for our existence. Cosmic Consciousness, anyone? This would certainly be a misunderstanding. Yuwayol does invite

*Yuwayol does invite you to consider a version of reality that may seem rather outlandish at first glance, but magical it isn't.*

you to consider a version of reality that may seem rather outlandish at first glance, but magical it isn't. I am not suggesting that Yuwayol means the existence of some ghostly soul which somehow flits between our physical bodies (though, at a long stretch, in certain respects, if you don't want to be too technical, I might admit that, figuratively, it may be useful to see it as such!).

In fact a good rule of thumb when reading any of what follows is this: if at any point you feel that you are being persuaded to adopt a spiritual, godly, magical or some other numinous belief, then one of two possibilities have occurred: (1) I have not been explaining myself properly (though I am doing my limited best), or (2) you have somehow misunderstood what I've written. Sometimes it is possible to interpret things two different ways. In such cases, should one interpretation seem in anyway supernatural, then it is probably safe to go with the other interpretation!

To be clear, whilst Yuwayol does make a remarkable claim (I certainly find it remarkable), it does not imply that the universe has a purpose, contains a spiritual realm, or has any supernatural dimension to it. These things, even if they are the case, provide nothing to the understanding of Yuwayol and are completely outside the remit of this book.

What can be argued is that Yuwayol does provide a moral imperative to our lives, perhaps arguably in a similar manner to the way that some religions do. This is discussed in later chapters, but for now it will suffice to say that, if you have a healthy moral compass already, Yuwayol should make very little difference to the direction in which it points.

## Yuwayol is not Science Either

Developments over the past hundred years or so have done a great deal to transform our understanding of the world and universe we live in. In particular, both Einstein's theories of relativity and our discoveries in the quantum realm have shaken many of our classical assumptions about how the laws of nature operate.

It is problematic that both of these branches of science are often non-intuitive and challenge our ideas of common sense. Relativity tells us that time and space are actually the same thing, and that there is no absolute time.

At least as bizarrely as this, quantum theory says that at the scale of the very tiny, objects such as electrons and photons can exist in different places at the same time, and that as well as being particles, they are waves.

This apparent abandonment of common sense by the scientific community does have the unfortunate consequence of encouraging the propagators of spurious notions to claim scientific support.

Quantum theory has been dubiously misused to back many a whacky idea, from spirituality, a time travelling Jesus, to quack medical practices.

So let me be very clear – Yuwayol does not pretend to be scientific. (However, to be fair, I do not think that it contradicts our current scientific understanding either.)

Science demands proof and I have none. Therefore I am no scientist. Yuwayol is an idea, a philosophy. Scientifically, it is speculation at best. I hope that when you have read this book you will be persuaded that there may be more to Yuwayol than you perhaps are now, but I will not be claiming to offer any scientific proof.

The reason I put the idea forward is because I think it an important one to consider.

Furthermore, I believe that Yuwayol makes sense of our existence in a way which no other philosophy does; that, in a broadly speaking way, it takes us as close to understanding our place in the universe as we are likely to get. In short, I feel Yuwayol is too profound and compelling an idea to just ignore.

*In short I feel Yuwayol is too profound and compelling an idea to just ignore.*

But whilst totally accepting that Yuwayol is not science nor does it ignore science altogether. In fact the chapter "*The Totaliverse and Yuwayol*" has a stab at putting forward some speculative views of the universe which neither contradicts Yuwayol or (as far

9

as I'm aware) current mainstream scientific understanding.

I do this in all humility, not because I think I know better than those experts working in this field, I clearly do not, but in order to try and demonstrate that Yuwayol is not as unlikely as it may initially seem to many of you.

## What's Coming

As previously mentioned, in the two chapters which follow on from this one, "*What am I? (The Question)*" and "*Why does anything exist? (The Question)?*" I will attempt to drive home the significance of these two ancient questions, their paradoxical nature, and why it is that they baffle us so much.

Following this, the chapter "*Levels of Understanding*" discusses the way in which historically our understanding of the world we live in changes depending on various factors, and speculates on whether an 'ultimate' level of understanding is achievable, and if it is then how would we know if we were to attain it.

I would be the first to admit that Yuwayol is not an immediately intuitive idea and is at odds with the way we perceive our world, however the chapter "*The Misleading Nature of Perception*" explains why it is not safe to presume that the way we perceive our world is the way the world actually is anyway, and suggests that much that we take for granted may actually be deceiving us.

At last, having laid the ground, I explain why I think Yuwayol should be treated seriously in the chapter "*The Yuwayol Perspective*" by linking back to the two big questions referred to earlier. A warning: an absolute answer to these questions is not given, however I hope to persuade you that Yuwayol somewhat compellingly removes their paradoxical nature.

As previously mentioned, the chapter "*The Totaliverse and Yuwayol*" should be treated as highly speculative and presents ideas about how Yuwayol might be applied, bearing in mind our current understanding of the cosmos we live in, to such phenomena as time, space, the big bang, entropy and the expanding universe. This may all seem a bit overwhelming to those of you not familiar with these topics, and probably lacking in rigor for those of you who are, but the basic idea of the chapter is to make the point that Yuwayol does not necessarily contradict anything that we so far know about how our universe works.

"*Getting Your Head Around Yuwayol*" is something of a general discussion of Yuwayol and aims to help you get used to the notion of what it means, and what it doesn't.

Earlier in this introduction I suggested that, under Yuwayol, morality is left unchanged for those with a healthy moral compass. Well that is my opinion anyway. The chapter "*Moral Imperatives*" argues for that opinion but looks at how Yuwayol might, or might not, persuade those of us with lower morals to behave better.

Yuwayol is not a religion, it's an idea. And though Yuwayol doesn't rule out god completely, the concept of Yuwayol and god don't sit very easily beside each other. But if you were to say "I believe in Yuwayol", would that make Yuwayol a religion? What is a religion anyway? The chapter "*Is Yuwayol a Religious Thing?*" chews it over.

And finally, what if Yuwayol is wrong? What if all things are different things much as we perceive them to be? Well, it turns out that I'm just as much you as I am me anyway. "*And if Yuwayol is Wrong?*" explains why.

**Talking About Your Self**

When making grand statements such as "you are me", "I am you", or "we are all the same thing", we need to be very careful what we actually mean by pronouns such are "you" and "me" etc. otherwise we're going to get pretty confused.

Think about your physical presence as you read this paragraph. This presence has a body which has grown over time from being an embryo in your mother's womb, to what it is now. It is a functioning organism which includes a face, a heart, a brain and all the other components that make up the human body. As your body has grown you have developed memories, characteristics, likes and dislikes and all the other things that make up a human personality.

And you have a name. I have a name also, my name is Secret Phil (actually not my real name but let's just assume it is). In this sense you, whatever your name might be, are not Secret Phil (probably!).

However, in this book we are referring to a more fundamental notion of "you" which encompasses and underpins this personality of yours, which you might think of as the deepest or essential "you". More of this in later chapters but this is the notion of "you" which Yuwayol says is also Secret Phil (me), as well as being you and as well as being everyone else.

For this reason, and from this point onwards, please bear in mind that in these pages the word "self" should be taken to be referring to this deeper meaning of "you", "me" or anyone else. So, for instance, when I refer to "you" I am doing so in the common manner, addressing you as your friends and family would tend to do. However, when referring to the essential "you" I will tend to use the term: "your self" (as opposed to the single word "yourself", please note).

> *...in these pages the word "self" should be taken to be referring to this deeper meaning of "you", "me" or anyone else.*

So, bearing in mind this precise definition of the word "self", I can accurately state that "Yuwayol claims that your self is the same entity as my self", whereas the statement "Yuwayol claims that you are me" might in certain contexts be misleading and cause confusion. Sticking with this convention will I think be preferable than trying to precisely explain what I'm saying every single time I use a pronoun.

So, what exactly is it we mean when we use words such as "you" or "me" anyway? Please read on…

# Chapter 2

# What am I? (The Question)

*"Being is the great explainer."*
**Henry David Thoreau**

### Who do You Think You Are?

How would you define yourself? I don't mean your name – your name can be changed and you would still be you, wouldn't you?

Nor am I asking you about your job or the things that you do in your life. If, tomorrow, you were to find that you could not do those things anymore, you would still be there, wouldn't you?

Or you may wish to define yourself according to the relationship you have with others around you, your family and friends perhaps. And this may well be as important to you as it is to me; but even alone, you are still there, aren't you?

*...bereft of all your possessions, you would still be you, don't you think?*

Similarly, your possessions, your clothes perhaps or your car, if you have one, might well say much about your character; but you could lose them all and, bereft of all your possessions, you would still be you, don't you think?

So what about the physical you – your body. Isn't that you?

Well we might think so, except that the body is constantly changing. First of all it typically changes in size and appearance, from conception where it exists as a single fertilized cell, to birth and constantly grows and develops throughout childhood, from there it goes through a metamorphosis during adolescence to emerge as a fully-fledged adult. After a decade or so at its peak, sadly perhaps, it slowly declines, eventually growing old before it withers and dies.

**Trigger's Broom**

But apart from this progression from womb to grave there is another way in which the body constantly changes. A nice analogy of this is to be found in a scene from the British TV comedy series *Only Fools and Horses*. One character in the show is a not so switched on character, nicknamed Trigger, who sweeps the streets for a living. In one episode, whilst drinking in the pub with his pals, he claims to be still using the same broom he was originally given when he started the job, "maintained it for 20 years," he tells his incredulous friends, "this old broom's had 17 new heads and 14 new handles in its time!"

In fact our bodies are a bit like Trigger's broom in that the parts that make them up are constantly replenished. Our body is made up of billions of many different types of cells. But these cells continually die to be replaced by others. The dust in the average household is mostly the product of dead skin, discarded from the body. Even the longest living of our cells are constantly replacing their components with fresher material.

So given all this constant changing, can we really say that our bodies are truly what make us definitively us?

**The Brain and Consciousness**

But perhaps we are being overtly materialistic. One very special part of the body is the brain. Now the human brain is often credited as being the most complex object that we know of. And perhaps the most remarkable thing about the brain is what it gives rise to - our thoughts, our emotions and our memories and desires and so on.

*...our thoughts, our emotions and our memories and desires and so on... Are not these the things which makes us what we are?*

Are not these the things which makes us what we are? It is along these lines that I think people in the modern world who have thought about it tend to think, and would feel inclined to define the notion of self.

The brain is indeed a remarkable thing and those who research how it works have made impressive progress in recent times trying to understanding it. But how the brain brings together all of its disparate activities, and all of the information fed into it, to create the coherent and single entity that we consider to be our selves, is far from understood.

19

This is not to say that we will not come to understand this process at some time in the future. But, however successful we may become in these endeavors, I'm not sure that this will bring us closer to answering the question we are actually asking.

Let me remind you, the question we are asking is "what am I?" I would argue that the tendency these days is to confuse this with the question "why am I conscious?", or even "how is consciousness possible?" Some cognitive scientists refer to this as "the hard problem".

Perhaps one day it can be solved, but if it is I think it would be a mistake to think it answers our question ("what am I?"). In order to illustrate, let's mess around with a couple of thought experiments.

### Beam me up Scotty, Twice!

It's a good bet that you will have heard of Star Trek. The definitive sci-fi TV series which originally aired in the late sixties and early seventies came up with a handy futuristic device for getting the Starship Enterprise's crew on and off planets. The 'transporter' was able to 'beam' objects, including people, from the Enterprise to wherever they wanted them to go.

On TV we would watch as our heroes vanished (accompanied by a sparkly light display and vaguely melodious sound effects) and then at the next moment they would re-materialize at their destination, miles away, ready for whatever adventure awaited them. Wonderful!

Of course this was all science fiction and present day technology couldn't even hope to emulate Captain Kirk and co. However, let's try and imagine how, inspired by Star Trek's transportation system, such a device might conceivably work at some point in our future.

When in use, our transporter would first have to make the passengers disappear from their initial location. And whilst doing so it would need to scan each and every atomic particle of their bodies (and their clothes I suppose), not to mention the state of each particle, such as temperature etc. and other helpful information like which direction the blood was flowing through their veins perhaps. All this information would need to be stored, and then used to reconstruct our traveler at their designated destination.

How this reconstruction might work is anyone's guess but let's just imagine that due to the genius of our future technologists, most likely inspired by Scotty (the Enterprise's highly put-upon chief engineer), a new living body, identical to the one that has just vanished is able to be successfully reconstructed.

Now the question is, although the person at the destination point is identical in all ways to the person who has just disappeared at the origin, and has the same memories, opinions and character traits, is this really the same person, or just a copy?

With this in mind, just how would you feel about using this method of transport?

To illustrate this further let's suppose that a malfunction occurred and, using the information stored in the initial scan, not one, but two people were generated at the destination, both of whom would be identical in every way to the original would be traveler, and therefore identical to each other. Let's say that you, not having considered these potential complications previously, were one of the (now two) passengers who had just been beamed.

*It would be strange indeed to meet an identical copy of yourself...*

It would be strange indeed to meet an identical copy of yourself, someone who had all the experiences that you had, the same tastes, the same fears.

No less strange would be the fact that they would be regarding you, not themselves, as the interloper; they would also probably see themselves, amongst other things, as the rightful owner of your bank account, and as the one who should rightfully live in your house, sleep in your bed and take your place within your family; and they would know all your secrets, just as you would know theirs.

So, you might be thinking, what does it matter? To have two copies of the same person might be socially embarrassing, and legally tricky, but you may conclude that this is still no reason to think that the self is anything other than the output of our brain.

But there is a reason why I am encouraging you to imagine that it is you who is the victim of our transporter mishap, because *only you* can know that *you* exist. I will expand more on this later, but for now let's consider a second thought experiment.

## What would it take for you to be not you?

A friend of mine, let's call him Fred, once told me how as a child his mother had related to him the story of how, when she had been a young girl, she and her family had almost migrated from the UK to Australia. The boat trip taking them to their new home was booked and they were all set for the journey but, due to the outbreak of war, the voyage was cancelled at the last moment.

Her concluding remark after telling Fred about this was something like, "just think, if we had gone to Australia you would be a little Australian boy now, instead of a little English boy."

Fred told me he had often puzzled over this in the following months after his mother had made the remark. He had still been young enough to trust that his mother was right in what she had said. But at the same time he had reasoned that if she had gone to live in Australia she would most likely have never have met his father, and so she would have married some other man. And if she had done that, and had had a child by the other man, then that child would not have been him, would it? Surely, he concluded, it would have been someone else, and he would therefore have never existed. Being a young boy, the thought had slightly confused and had strangely unsettled him for a time.

I think most of us can agree with Fred, that his mother's child, born by a different father in a different country would probably not have been Fred. But maybe the problem goes a bit deeper than that. Suppose the difference between Fred and this hypothetical other child, let's call him Fred2, was more subtle.

Let's go back to the occasion when Fred was actually conceived, and let's imagine that something different from what did happen took place. Suppose a different sperm had fertilized the mother's egg. Would the resulting offspring be Fred, or would it be Fred2?

Or let's imagine that the same sperm had fertilised the egg after all, but by a fluke a radioactive particle had passed through his father's body, making a change to the sperm's DNA. Would Fred then still be Fred?

To answer this we need to know what it is that makes us actually us. And yet no sensible answer seems to be available.

As I have already pointed out, we, all of us, change all the time. When we move from child to adult the change is particularly dramatic. Our bodies change, as do our thoughts and feelings, but as adults we still think of our childhood as something which happened to our self. But is that correct?

Once the child has gone perhaps he or she is gone forever. Perhaps we are newcomers on the scene, inheriting the memories of a child that no longer exists, essentially a different person. And perhaps as we get older still we become another person yet again. So at what point do we stop being who we are?

> *Perhaps we are newcomers on the scene, inheriting the memories of a child that no longer exists....*

## Maybe there is no real me or you, just the illusion of self

At this point it is tempting to wonder if we are just chasing a phantom. Maybe the 'you' I am talking about doesn't really exist at all.

In the same way that a motorbike can produce the phenomenon of speed, so our brains produce the phenomenon of consciousness. And when that consciousness becomes aware of its own existence then isn't it bound to arrive at the illusory belief that its self is something real and tangible?

Quite a few people these days think upon those lines, and it stems, I guess, from the modern approach we have towards thinking about these things. We observe, we make notes, we formulate theories, we test our theories with experiments, and then, if our theories match the results of our experiments, we draw conclusions.

> *...it is only when you look at your own self, and experience your own existence, that the puzzle really becomes apparent.*

This is the scientific approach, and I would certainly not want to knock it. However, you can observe as many test subjects as you wish, allow as many dolphins and chimpanzees to play with mirrors as you want, conduct all the brain scans that you can, but it is only when you look at your own self, and experience your own existence, that the puzzle really becomes apparent.

People sometimes talk about the elephant in the room, referring to something obvious which nevertheless nobody seems to acknowledge. But the Self must be the bulkiest and most awkward elephant in the scientific room today, yet frustratingly I cannot think of a single scientific experiment that is ever going to test the hypothesis that you really are you.

And yet in spite of all this, *you know that you are you*; if you think about it, it is the only thing that you actually do know for sure. Everything else that you think you know might be an illusion, but you aren't. I'm not saying (at least not now) that this essential you is important or even substantial, but the thing you call yourself is definitely real, even though no one else can ever really know it.

I say that you are definitely real, but then again I have to concede that, from my perspective, there is always the possibility that you aren't real at all; that you are a kind of zombie aware of yourself only in the sense that a computer virus can refer to, and duplicate, itself.

The curious thing here is that if you are a zombie there is no way that I would be able to tell. I could ask you if you were real, and you would look upon yourself and say to me, "of course". Similarly, there is no way you can know that I am not such a zombie.

And so I say to you (and I'm sorry if I seem to be going on about this, frustratingly I can only point this out, only you can be aware of it), look to your own self and ask yourself: am I real? There is only really one answer: of course you are!

And that answer is a real puzzle. Not least because whenever we go looking for the self it doesn't seem to be there. Seemingly, the deeper we look into ourselves our self disappears from view. Perhaps we are looking in the wrong direction.

*Seemingly, the deeper we look into ourselves our self disappears from view.*

Also, doesn't it strike you as astonishing that of all the people who have ever lived, and the billions times more people who might have lived, but didn't (such as the Australian Fred2), doesn't it seem remarkable that you are – you? Suppose you had never lived?

How come you exist at all? Come to mention it, why does anything exist?

# Chapter 3

# Why Does Anything Exist?
# (The Question)

*"It is not how things are in the world that is mystical, but that it exists."*
**Ludwig Wittgenstein**

### Nothing

I want you to try and imagine that nothing exists. Do it now – take as much time as you like, close your eyes if it helps and try and imagine that there is nothing, that there never was anything, never will be anything; that time and space, you or I, or anything else has simply never been there and never will be there...

...think about what you just imagined. If you are like me it is unlikely that you were totally able to conceive of a complete absence of anything. I tend to visualize blackness, emptiness, a bleak place where nothing happens. But actually that is not nothing either. If we have nothing, then there is not even blackness, for there can be no concept of black. Nor can there be emptiness, for there is no space that can be empty. And there is no bleak place either, for there is no place to be bleak, and nothing can happen for there is no time for it to happen in! There is *nothing*.

I'm not sure that it is even possible to truly imagine nothing. But I think the exercise is worthwhile because, if you think about it, isn't that how it should

> **_Surely, by default nothing should exist?_**

be? Surely, by default nothing should exist? It is difficult to appreciate something existing when there seems to be no reason of why it is there.

We are used to this notion in our everyday lives; things exist because they were somehow produced by other things.

This is certainly the case for the things in our everyday existence. This paragraph that you are reading, for instance, is there because I wrote it and made it available for you to read. A puddle of water on the ground is likely to be there because it has rained. A tree exists because it grew from a seed having been nourished by the earth, sun and water. The teapot in my cupboard is there because someone made it and I bought it from a shop and put it there. We are used to this notion in our everyday lives; things exist because they were somehow produced by other things.

> **_We are used to this notion in our everyday lives; things exist because they were somehow produced by other things._**

But where, you might well ask, did I, who wrote the (previous) paragraph, come from? Where did the rain come from which filled the puddle? How do you account for the earth, the sun and the water which made the seed grow? Whether or not we know the answer to these questions we generally presume that there is an answer out there. But when we ask the question "why does anything exist?" we have a problem, because suddenly we don't have the something for the anything to have come from! And so there should be nothing.

And yet here we are!

The problem is as old as human history and there have been numerous attempts at answering it, most notably from the domains of philosophy, religion and science. These explanations can roughly divided into three categories, each with their own problems, which I define as (1) supernatural, (2) finite universe, and (3) infinite universe. Let's take a look at each.

## Supernatural – was everything created by a supreme being?

The notion of god, or gods, is very popular. Belief in these deities take varying forms, which we usually refer to as religions, and many of these religions have their own particular stories of creation to tell. For now I will focus on the belief, shared by many religions, in a single all powerful god. However, I think the same arguments used can be applied just as easily to the belief of many gods.

> *For many people, this is the answer to our question: Things exist because God created them. Easy!*

For many people, this is the simple answer to our question: Things exist because God created them. Easy!

But of course not everyone believes in God, and for those people there is a fairly obvious repost to the idea that God created everything: OK, they can say, so if God created everything, who (or what) created God?

In return the believer has an answer which invariably goes something along the lines of: Don't be silly, God wasn't created, he has always been there, he always will be. You don't create God because God is omnipotent – he *is*. Most believers find this argument to be entirely reasonable and to a certain extent I can sympathize.

> *When we say "everything"... then, if he exists, "everything" has surely got to include God.*

The problem is that in taking this view (that God created everything) the question of why everything, rather than nothing, exists has not actually been answered. When we say "everything" (certainly for the purposes of this question) then, if he exists, "everything" has surely got to include God. It may well be that having a belief in God, and accepting that everything we see is something which he has created, satisfies you for reasons of faith, but our original question, why does anything exist (including God), has been set aside.

Another way of approaching this is to think of it this way – if you were God, wouldn't you wonder why you were here?

Now at this point you might argue that God, in his omnipotence, would be able to comprehend the explanation of his existence; after all he (or she?) is not like us, he comprehends all things including things that we are unequipped to grasp. But by making this argument you have subtly deflected us from the original question.

We are not trying to determine if God's existence is feasible, or what he might or might not comprehend. We want to understand why *anything* exists, God (if he does indeed exist) included.

This may well be something which God might know but we certainly don't. And so we have found a new way of saying that we do not know the answer to the question of why anything exists. For this reason, irrespective of God's existence, saying that God created everything does not answer the question of 'why does anything exist'. In fact it seems to conclude that we should not bother asking the question at all.

So to summarize – the existence of god (or for that matter, gods) has not answered the question of why anything, rather than nothing, exists.

### Finite Universe – Big Bang and Big Crunch

A finite universe is one which has boundaries and is therefore limited in size and scope. A good example of this is provided by the Big Bang theory incorporating the idea of the Big Crunch.

Modern cosmologists mostly hold the view that the universe we see around us began as a singularity around 13,750,000,000 years (a long time) ago. A singularity has no size; it is a point, like the very tip of a perfectly sharp needle. This point contained everything, in a super condensed form, that is in the universe today and has since expanded to its current vast proportions.

This is the Big Bang theory and, as remarkable as it may sound, the evidence for it is extremely compelling (for an understanding of why modern cosmologists are certain that the Big Bang did actually take place I would recommend the comprehensively titled book *Big Bang: The Most Important Scientific Discovery of All Time and Why You Need to Know About It* by Simon Singh).

The term 'Big Bang' itself is a bit crude but it describes those opening few moments when the universe seemingly burst from nowhere.

*...trying to comprehend the universe prior to the Big Bang is generally thought to be futile.*

Before the Big Bang, it is said, there was nothing; not even time existed. Space did not exist either; space came about during the Big Bang, expanded, and continues to expand to this day. In fact trying to comprehend the universe prior to the Big Bang is generally thought to be futile; there simply was no pre-Big Bang. Imagining such a thing would be no different from the exercise we performed at the beginning of this article when we tried to imagine nothing.

(Now, for completeness, I should mention that a number of cosmologists, whilst they would not dispute that the Big Bang happened, do question the hypothesis that there was nothing prior to the Big Bang. There are a number of suggestions of what might have come before, but to my mind they mostly fall into the category of Infinite Universes, which I cover below, or simply provide no answer as to where everything came from (I dare say there are theories I have never heard of, but for the purposes of this argument I think we can bypass them). So for now I would like to treat the Big Bang theory as it is, I think, still mostly understood – that it was the origin of everything.)

The obvious question here is – Why should there have been a Big Bang at all? Why would this sudden and dramatic moment of creation just happen, seemingly for no reason?

Addressing this, in his excellent book *A Brief History of Time* the famous theoretical physicist, Stephen Hawking referred to the idea that the universe may one day stop expanding, and under the power of gravity, it would begin to shrink, eventually collapsing back in on itself in a cosmic Big Crunch, ending, just as it began, as a singularity.

At the Big Crunch all time and space would end, just as it had begun with the Big Bang. There will be no space or time after the Big Crunch, just as there was no space or time before the Big Bang.

Professor Hawking invites us to think about this speculative history of the universe, and likens it to a spherical ball. If you draw a line between two opposite points of the ball's surface, through the center of the ball, then this line represents the time line of the universe, the Big Bang at one end and the Big Crunch at the other. The surface of the ball represents the universe expanding from the Big Bang and then shrinking back to its original size at the Big Crunch.

Now, if you stand back and look at a sphere, you will not perceive a beginning or an end, just a sphere, with all points on the surface of that sphere being equal. Hawking's argument is that this means that there is no need, if you take this view (actually what you should be imagining is a kind of four dimensional sphere, which is easier said than done, but to be fair that does not affect the point being made) then there is no longer any requirement for a moment of creation, and therefore there is no need to worry about how the universe came to be.

Well (hesitant though I am to differ from Professor Hawking, for whom I have the greatest regard), I'm afraid I can't really go along with this. That isn't to say I don't think there was a Big Bang, or even that there won't be a big crunch. What I mean is that, however complete and consistent the Big Bang theory is, I can't accept it as an explanation of why there is something (the universe) rather than nothing.

If we accept that the dimensions of space and time of the whole universe throughout its history can be likened to a kind of self-contained ball (which for all I know it can be), then we are still left with the question, as we are with any such finite universe theories – how is it that, instead of nothing, this great and complex ball just happens to be there?

### Infinite Universe – Tortoises!

There is a story which may or may not be based on a true event. It is quite a well-known story and there are various versions, so I hope you will forgive me for completely inventing one of my own:

> *A young woman, who had recently completed a university degree, had taken the time to embark on a Himalayan trekking holiday in Nepal. During her journey she became good friends with the Sherpa who guided her. One day, making conversation, he queried her about her degree and what subject she had worked on. She told him that she had studied cosmology.*
>
> *"What is cosmology?" he asked her.*
>
> *"It is the study of the universe," she replied.*
>
> *"Indeed?" he said, impressed. "That is a very big subject."*
>
> *"Yes," she agreed, laughing. "I have to agree."*

"And did they teach you about where the universe came from?" queried the Sherpa.

So she spent a few minutes explaining to him about the Big Bang theory, and told him that was how most scientists currently believed the universe began.

The Sherpa listened politely and with much interest. "That is very fascinating," he told her. "But my Grandmother would not agree."

"Your Grandmother!" said the young woman, a little taken aback. "Really?"

"Yes," said the Sherpa. "My Grandmother says that the world is flat and sits upon the back of a giant tortoise."

"I see," said the young woman, both amused and very intrigued. "And what about the tortoise, does it have anything to stand on?"

"That is a good question," said the Sherpa. "But I don't know the answer. Perhaps you should ask her yourself." He then explained that at the end of the trek they would be able to pass the roundhouse (a traditional type of dwelling in Nepal) where his Grandmother stayed. He was sure, he explained, that his Grandmother would be delighted to meet her and answer her questions.

And so it was that the young woman, a week or so later, entered the roundhouse where the Sherpa's Grandmother lived. After an amicable introduction the Sherpa mentioned to his Grandmother about the

40

*tortoise which supported the world upon its back. The Grandma nodded.*

*"So what our guest would like to know,"* *asked the Sherpa, "is upon what is it that the tortoise stands?"*

*"Why," said the Grandma. "It stands upon the ground of its own world."*

*"So where," asked the student, "is that world?"*

*Grandma laughed. "That world sits upon the back of another tortoise."*

*"And that tortoise...?"*

*"...lives on a world on the back of another tortoise," Grandma told them. With bright eyes she leaned towards her younger companions as they stared back at her, puzzled, and she confided to them, "You see, it is tortoises all the way down!"*

The notion of a presumably infinite number of, ever increasingly large, tortoises all supporting the worlds above them seems rather comical (or it does to me!). But I'm not sure it doesn't illustrate the oddity of any other theory on offer.

The 'tortoises all the way down' theory is what we might describe as an infinite universe theory, which can be summarized as suggesting that the universe goes on forever in time and/or at least one dimension of space.

But we are still left with the question – how come? The idea of an infinite chain of big bangs might not raise a smile in the same way that infinite tortoises might, but is it fundamentally any less preposterous?

Or the idea of an infinite set of universes, spanning infinite time? Again, how come? Why should such a thing be? Because it just is, you might say. But if you do, you are simply turning your back on our original question – why does anything exist?

## Can Quantum Physics Give us an Answer?

Some quantum scientists have interesting things to say on this topic. As an example, a good relatively readable book on this subject is *A Universe from Nothing* by Lawrence M Krauss.

The claim (I'm generalizing to a very large extent here but hope I'm not being unfair) is that an understanding of quantum theory itself allows us to understand why there is something rather than nothing. Everything in the universe is energy. There is mass also, but mass can be thought of as a form of energy packed into a small area, as Einstein discovered. The argument goes that the universe contains negative, as well as positive energy and (I hope I'm not being too simplistic here) it seems that there is good reason to believe that the positive and the negative energy might well be of quantities that rule each another out.

This being the case then there is a sum total of zero energy in the universe, which means, at least from this perspective, that the universe might as well not be there – it might well be nothing, and that's why it was able to come from nothing in the first place – nothing to nothing, if you like.

Quantum theory tells us that what we think of as empty space is in fact teeming with virtual particles, positive and negative, which pop in and out of existence all the time, annihilating each other in the process (as remarkable and unlikely as this sounds, quantum theory, which happens to be the most experimentally proven theory ever, relies on this idea for its existence).

So, the argument goes, the entirety of the universe is allowed to exist from nothing so long as the positive and negative energies that make it up cancel each other out – the entirety of the universe could be thought of as a sort of grand quantum fluctuation.

*...the entirety of the universe could be thought of as a sort of grand quantum fluctuation.*

Well this is all very fascinating, and I'm very happy to go along with it. (In fact, as we shall see Yuwayol is comfortable with the notion that in a certain context the universe might well be thought of as not really being there!). However I'm not convinced it cuts the mustard so far as answering our question goes.

If quantum rules allow something to come from nothing, or perhaps to put it more pointedly, they do not allow nothing to exist, then it seems to me that we are left with the question of where these quantum rules emerge from. As a set of rules they may not be a physical thing, but they are most certainly a *thing*. And as such they do seem a bit arbitrary to say the least to simply exist in their own right.

## And Yet Here We Are

We know for sure that we are here (as discussed in chapter – "*What am I? (The Question)*"). So we know for sure that there is something rather than nothing. We just don't understand how that should be.

> *...there seems to be no explanation for why anything exists that would satisfy us even in theory.*

Actually it's worse than that, there seems to be no explanation for why anything exists that would satisfy us *even in theory*. And I should warn you once again, that no answer to this question is forthcoming (sorry, I hope you weren't expecting one!), however I think we may well get as close as we can do.

The purpose of this chapter is simply to help you the reader understand and appreciate our conundrum. It is as we contemplate this, together with our discussion of that other big question (What am I?) that we will begin to make the case for Yuwayol.

But before we do this we need to consider the limitations that, as humans, we have when we observe the world we live in. We shall do that in our next two chapters, "*Levels of Understanding*" and "*The Misleading Nature of Perception*".

# Chapter 4

# Levels of Understanding

*"One is always a long way from solving a problem until one actually has the answer."*
**Stephen Hawking**

*"Any sufficiently advanced technology is indistinguishable from magic."*
**Arthur C Clarke**

*"If I have seen further than others, it is by standing upon the shoulders of giants."*
**Isaac Newton**

### Of Humans and Ants

Picture the scene:

> *A colony of ants busy themselves around their nest, building and foraging and generally doing what ants do, much as they have always done. A man stands over them, a tube of fine white powder in his hand. He is concerned about the effect the insects are having in his garden. He puts on his spectacles and frowns as he reads the instructions on the side of the tube. "Apply powder to the main entrances of the wasp or ant nest," he reads. "The powder will attach itself to the insects' bodies as they enter the nest and will be distributed throughout the colony." Satisfied, the man puts away his*

*spectacles; he crouches down over the nest and uses his thumb to flip the lid from the tube.*

*Meanwhile the ants continue to do what ants do, much as they have always done. There is no way they can know that, by the time the sun rises the next day, they will all be dead...*

Many have considered it appropriate to compare humans with ants, and for various reasons. But usually this is in the context of making a point about scale – comparing ants on a hill with humans in the city perhaps. But the point I would like to make about the scenario above, is to do with what, in this book, is something I will be referring to as 'levels of understanding'.

We humans in a city would quickly spot some gigantic creature if it were to stoop over us with some murderous intent. If it were to lay down poison at certain strategic locations, we might even be able to take evasive action and minimize the scale of the disaster. Heck, with the weapons we have at our disposal we might even be able to take the giant on!

I digress.

Rather, what I think the scenario above really teaches us about is how different the world, or the universe for that matter, looks depending on your understanding of it, and your status within it.

Ants have been around for many millions of years; however, humans with plastic tubes of ant toxic chemicals have not been around for very long. Evolution has not trained ants to deal with such situations. If you stand over their nest they are oblivious to your existence, however large you might be. As remarkable as the internal workings of an ant colony might be, individual ants are relatively simple creatures. They live by scent, they follow trails, left by their fellows, and perform tasks in a pretty automated way.

Humans, on the other hand live with a very different level of understanding, and it is important to realize that each new level of understanding presents us

*...each new level of understanding presents us with a very different way of seeing the world we live in.*

with a very different way of seeing the world we live in. We are aware of ants in a way they could never be aware of us, we can observe and study their habitats and behavior, and we can test their genes and categorize them, even work out their ancestral location within the animal kingdom.

## Magic and Science

Humankind's level of understanding, when it comes to understanding the world and universe it lives in, has expanded quite dramatically over recent centuries. It is not that we are cleverer than our forebears, rather we have discovered much that has given us a very different perspective on things.

Let's take comets for an example. Haley's comet has made repeated visits to our night skies over the course of human history and has been greeted with concern in ages past. When you believe that the heavens are the abode of god, or gods, then such a new star, with its spectacular glowing tail, might well be regarded as a worrying omen.

In one of its more famous visits in 1066 the comet was greeted at the time by the English chronicler, Eilmer of Malmesbury, with these words – "You've come, have you? ... You've come, you source of tears to many mothers, you evil. I hate you! It is long since I saw you; but as I see you now you are much more terrible, for I see you brandishing the downfall of my country. I hate you!"

Weeks later the king of England, Harold II, was killed at the Battle of Hastings and replaced as king by his victor, William, the invading Duke of Normandy.

That a comet might portend the demise or ascendancy of kings and empires was not considered unusual in times gone by. History is littered with the examples of such. It would be interesting then to know what Eilmer of Malmesbury would have made of things recently when a spacecraft, built in part by his people's descendants, managed to land a probe on a comet some 500,000,000 kilometers from earth.

> *…we know a great deal more about comets than we did in the eleventh century.*

Due to such endeavors we know a great deal more about comets than we did in the eleventh century. We know, for instance, that they formed at the infancy of our solar system, along with the earth, some 4.5 billion years ago; they are composed mostly of ice and rock; they orbit the sun in huge elliptical orbits; and their 'tails' are formed by escaping dust illuminated by solar winds from the sun (they always point away from the sun rather than from the direction the comet is travelling).

Remarkable things comets, but it is doubtful that they have any bearing on our squabbles here on Earth, and probably not deserving of Eilmer's vitriol.

And of course it isn't just comets that we have improved our knowledge of. If time travel were possible then any reasonably well educated person could probably amaze and astound those who lived in 1066 with their knowledge of the world, assuming they believed you and didn't execute you for being a heretic. And you might well be taken for a sage or magician, especially if you were able to demonstrate a few twenty-first century gadgets to them.

But before we get too wrapped up in how clever we have become, it is worth pondering what new levels of understanding might be achieved by our descendants in the millennia to come, should our race

51

survive that long. If *they* were to visit *us*, would they smile politely at our flashy smart phones? Or be underwhelmed by our fastest computers? Snigger amongst themselves at the naivety of our technology and science? For now, we can only imagine what seeming magic they might dazzle us with.

## Limits of Understanding

So, if the human race were given enough time, might we eventually reach a level of understanding that cannot be improved on? Is there, in other words, a highest level of understanding, an ultimate understanding of truth?

*Is there, in other words, a highest level of understanding, an ultimate understanding of truth?*

It is important to note here that I am not discussing the possibility of us knowing everything. By a 'highest level of understanding' I am talking *about a complete understanding about fundamentally what the universe is and how it works*.

Interestingly there was a point in our history when it was thought by many that we were close to such a level of understanding concerning the rules which the physical universe follows. At the dawn of the twentieth century Isaac Newton's laws reigned supreme, and seemed to account for almost everything. We knew about atoms and it was to be expected that these small objects most likely followed Newton's laws albeit at a miniscule level.

There were a few loose ends, it was thought, which needed to be tidied up, such as the nature of light and magnetism. But it was by studying the loose ends that we were led on a bewildering re-writing of the rules of physics. Newton's laws were overturned and the pioneers of relativity and quantum mechanics set us on a new road of discovery which we still travel today.

> *...it was by studying the loose ends that we were led on a bewildering re-writing of the rules of physics.*

So, bearing in mind the misplaced confidence of some of our early twentieth century scientists, one question which might be worth considering is how we would even know that an ultimate level of understanding had been reached.

But before considering this question it might be worth looking at a couple of reasons why reaching such a level of understanding might not be attainable even in theory.

One reason might be that the highest level of understanding is not in our reach and never can be.

There are various ways in which this might be the case, but by way of an example let's consider this: It has been mooted that the world we live in is a kind of computer simulation. Without going into the details of this idea (for more information you can start by searching for 'simulation hypotheses' on the internet) this theory basically suggests that our world has been developed, as a piece of advanced computer software, by some highly technologically capable 'people'. Why they would do this is largely immaterial for this discussion; perhaps we are part of some scientific experiment, or maybe we are involved in a computer game.

No matter. The point is that if this scenario were to be the truth (there are some who take the suggestion perfectly seriously) it would be clear that there is a distinct limit on what level of understanding we can achieve. Locked in the confines of some higher dimensional memory chip it is difficult to see how we could learn anything about how this memory chip was manufactured, the 'people' who manufactured it, or anything about the world in which those 'people' live. We would be as clueless about that world as a character in a computer game would be about ours.

Physics in that place might follow a completely different set of laws to the ones followed in our universe (which presumably have been programmed into the simulation software), and without being able to study those rules, how could we ever hope to understand them.

We could of course always study the simulated world we live in (provided our programmers continue to keep the program running and no one pulls the plug out!), but our ability to probe the workings of reality would be forever hobbled.

A second constraint on our understanding might be that we are simply not clever enough, that our brains are just not up to the job of understanding any of the ultimate significant truths that may be out there, in much the same way that the ants described above could have no way of appreciating the manufacture of tubes of deadly white powder.

*A second constraint on our understanding might be that we are simply not clever enough...*

When you think about it this is an entirely feasible idea. The laws of evolution tend to give rise to creatures which are adapted to the environment they live in. And what can we say about the environment we humans have evolved in?

Well, as big and as wonderful as the earth might be to us, it would seem to be no more than an insignificant lump of rock utterly lost in the unimaginable expanses of space and time. Isn't it hubris to think that we might ever be capable of understanding everything?

Perhaps, but it might also be worth pointing out that, as small as we are, we have come an impressively long way up to now.

But for the moment it must be admitted that there is no way of knowing if we can ever achieve the highest level of understanding, but if we are ever to get there it might be worth returning to the question posed a few paragraphs back: if we were to attain the highest level, how would we know?

## On the Top Floor

Imagine a tower block of many floors. And let's let each floor level metaphorically represent a level of understanding; so the higher up you go the higher the level of understanding you have.

In this tower block you have to figure out how to get to the next floor. There are no windows and there is no central easy to follow stairway, the way to the next floor up is rarely obvious and is by secret hidden stairways, and doors with locks which can only be opened by hidden keys.

In this tower block the ants live quite low down on a floor not so far up from the basement. (I hope you are not forming the impression that I have anything against ants, in fact I think they are extraordinarily fascinating creatures!)

On a reasonable number of floors higher than the ants live our human ancestors who at some point discovered how to make use of fire for cooking and keeping warm.

In order to light their fires they learnt that they could produce sparks by knocking pieces of flint together. Now whilst they no doubt developed some expertise doing this (probably much greater than most of us living today), there is no way they could have understood *why* it was that knocking two pieces of rock together produced sparks.

They may well have developed ideas or theories why such a fortuitous phenomenon should take place, perhaps involving the odd god or so, but without even a basic understanding of physics they could never have understood what was going on as they struck their flints to create fire during those cool prehistoric nights.

But here in the twenty-first century we have (or at least those amongst us who have studied the physics) the means to understand more clearly what goes on at the atomic scale within those rocks.

We know much more about how energy works, and about photons, the carriers of energy, and of how they can be released when an electron moves from one 'orbit' to another within an atom. And so we occupy a higher level in our imaginary tower block than do our ancestors.

But let us not get too smug. Having a higher level of understanding does not mean we are smarter than those on the lower floors. We stand on the shoulders of giants, as Isaac Newton once put it.

It is true that we have our lasers and our silicon chips and we have these because of our understanding of quantum physics. In the quantum realm we can make extraordinary calculations about what should happen at the atomic scale and find them to be exquisitely accurate.

And yet for all that, we seem to be miles away from understanding what it all means. Like our forefathers banging rocks together to produce fire, we make calculations and design our impressive devices without a genuine understanding of why what we do actually works.

> *We seem to be miles away from understanding what it all means.*

And so we know there are higher floors to be discovered. And as we search for a way up, to the next level of understanding, it might occur to us that as high as we've come already, following quite literarily what might be termed a bottom-up approach, we have no way of knowing how many floors the building actually has.

But what if we were to reach the very top floor? Would we know? It may well be that we would not, but I suspect that the top floor would be somewhat different from any of the previous levels.

Rather than being like any of the lower floors I fancy that what would happen is that we would finally emerge onto the very roof of the entire building where, blinking in the sunlight, we would be under no doubt that there was no further to go, and that we were indeed on the top level of the building.

Similarly, if a highest level of understanding is feasible, and if our descendants were to ever reach it, I fancy that it would be self-evident. Yuwayol has something to say about what the main characteristic of that highest level of understanding would be. Due to reasoning to be presented later, Yuwayol says that all things are fundamentally the same thing; that all things are one.

If Yuwayol is correct, then at the highest level of understanding, this oneness would be demonstrably self-evident and proven.

# Chapter 5

# The Misleading Nature of Perception

*"The greatest obstacle to discovery is not ignorance - it is the illusion of knowledge."*
**Daniel J. Boorstin**

*"Reality is merely an illusion, albeit a very persistent one."*
**Albert Einstein**

### The World You Live In

The human brain is the most complex object that we know of.

It may well be that you have heard that said before, so how about this: So amazing is the brain that it creates the remarkable world, in all its diversity, in which you live.

> *So amazing is the brain that it creates the remarkable world, in all its diversity, in which you live.*

"Er… hang on a minute," you may be thinking. "Yeah, the brain is pretty amazing, but it sits inside my skull. There is no way it can possibly have created the world I live in. The world I live in gets on just fine without me."

And yet, I insist, my claim is true. *You*, the real you, the you consciously reading this book, live in a simulated reality which has been created by your brain for *you* to exist in.

Everyone you know, every memory that you have, every street you have walked down, every meal you have eaten; the place you live, the sky you see, the music you hear, this book that you are now reading. All of these things have been generated in your brain for you to experience. If this sounds like nonsense to you then please take a pause, stay with me, and think about it.

Now what I'm not saying is that the world on the outside of your skull is not real.

The people you know are really out there (well, I can't actually prove that to you but they probably are, I hope), but to what extent they are like the people you perceive is somewhat open to question. In fact it is equally open to question whether anything you perceive is actually like it really is. In order to illustrate, let us take something which most of us very commonly experience as an example, let us take the color blue.

I personally like the color blue. Actually, I like all colors, but I find blue particularly pleasant somehow. Now let's say that we have a blind friend. Someone who has never seen, and so has never had any experience of, the color blue. Wishing to share with her the pleasure I get in looking at the color blue, it would be a nice thing to describe it to her. But how do I do that? Well, that's the thing; I can't.

When you stare at something blue, let's take a cloudless sky on a sunny day for an example, certain cells in the back of the eye, which are sensitive to a certain wavelengths of light, respond and send signals through the optic nerve to the brain. Note that this signal is electrical in nature and does not consist in any way of the light which triggered it.

What the brain does next is nothing short of remarkable.

Based on such information as the direction the eye was pointing when the signal was received, and the degree to which the eye was focused on what it was looking at, and using the information simultaneously coming in from the other eye, the brain 'paints' for the mind, for you, an image of the sky which, even more incredibly, is 'painted' blue.

The important thing to infer from this is that the blue which you experience has been generated by the brain. Blue, as you experience it, does not actually exist anywhere except in your brain (and maybe other peoples).

*Blue, as you experience It, does not actually exist anywhere except in your brain.*

Another way to think about this is to say that the brain has invented the color blue in order to represent something 'out there', namely a certain frequency of electromagnetic radiation (otherwise known as light). And this is why our blind friend can never know what it is like to see blue. If her brain has never painted the color blue for her, then she can never know how it 'looks'.

Indeed even our sighted friends may not know what the color blue is like for you, because there is no way of being sure that the blue you see, when you stare at a blue sky on a sunny day, is the same as the blue that they see when they are looking at the same thing. Their brains may very well be generating different colors altogether, colors you can't begin to imagine, and you would never know because you are using your own brain, and not theirs.

**Virtual Reality**

And it doesn't end with the color blue.

All the different colors we see are generated in the brain in response to any light wave which falls within the visible spectrum, and it does this in order to make sense of the signals it receives from the eye (there are three types of light sensitive cell in the eye, each responding to a differing range of wave lengths; it is the strength of signal from each type of cell that prompts the brain to generate all the colors that we see).

And it doesn't end with the sense of vision either. All of our experiences, it turns out, are generated in the brain. Let us take sound as a further example.

There is no sound in the world outside our brain, not as we experience it. Instead of light waves which our eyes detect, our ears are sensitive to vibrations in the air. These vibrations make the eardrum vibrate and the delicate mechanisms within the ear generate differing electric signals to send to the brain. And the brain in turn reads these signals and based on them it produces what we experience as noise.

So when someone plunks the C sharp key on a piano keyboard they are not, strictly speaking, producing a sound, however lovely. They are simply producing vibrations of a particular frequency in the air around them. It is only when the ear informs the brain of this event that, once again by virtue of the brain, the experience of sound is created. There is no C sharp, just the C sharp invented by the brain.

Nor does it just end with sound and vision. All of our senses can be described in this way.

I was taught at school that we have five senses. Sound and vision were two of them and in addition I learnt that smell, taste and touch should be added to the list. In fact my teachers understated our sensual activity. To date there are in fact up to twenty-one recognized senses.

For instance there is our sense of balance, which informs us which way up we are, as well as the sense of proprioception, which allows you to know where your various body parts are (try shutting your eyes and touching your nose with your finger, it is thanks to proprioception that you are able to do that).

And all we have said about vision and sight apply equally for all the sensations and perceptions that all of these senses generate. They are all, through various means, picking up signals in the world around them, including our own bodies, they then interpret and communicate information about those signals via our nervous system to our brain, and our brain then takes them and produces the cacophony of sensations that we experience.

But it doesn't just end with our senses. The brain performs yet another exceptional trick.

Because somehow it manages to take all these diverse signals, from our eyes, our ears, our touch and so on, and fuses them all into a single conscious whole, which it binds into a series of flowing moments which we experience as our ongoing lives. A virtual reality if you will, brought into being by the remarkable 1.5kg lump of organic matter (mostly fat) within our skulls.

*A virtual reality if you will, brought into being by the remarkable 1.5kg lump of organic matter (mostly fat) within our skulls.*

Impressive hey? It is indeed a most wonderful trick. Incidentally, quite how our brains achieves this is not very well understood, figuring it out is something cognitive scientists often refer to as 'The Hard Problem'.

Of course this is not something we spend a lot of time thinking about as we go about our lives. We generally take it for granted that what we perceive is representative of the big wide world that our brains physically inhabit. We assume that our own personal virtual reality reflects what really is 'out there'.

But to what extent is that a safe assumption?

### Reality or Illusion

Just how much does this virtual reality world, conjured up in our heads, actually correspond with the real world?

Well, in many respects, it would seem to correspond very well indeed. The fact that you can walk quite happily down the street, negotiating traffic, people and other objects, typically without incurring any harm, would suggest that your brain's virtual reality serves you quite competently.

If there's a lamppost to be avoided, and you are paying reasonable attention, and you are sober, you don't walk into it!

But whilst we seem to be able to navigate around quite well, there is plenty of evidence to suggest that we can, at least in many subtle ways, be deceived.

This is because, as impressive as it is, the brain is not perfect. For the sake of efficiency it makes quite a lot of assumptions and takes shortcuts, all of which can mislead us. In other words the brain can often present the world, not as our senses report it, but as it presumes the world to be. And sometimes we can catch it in the act of getting things wrong.

*...the brain can often present the world, not as our senses report it, but as it presumes the world to be.*

For a good example check out the internet for a demonstration of the McGurk effect (there's a good one on YouTube – search for 'Try the McGurk Effect! - Horizon'. This demonstrates how what you see with your eyes can actually change what you are hearing. The brain is using what our eyes are detecting and overriding the signals coming from the ear, and we actually hear a sound that is not as it really is.

There are, in fact, numerous other examples I could show you which illustrate the way our virtual reality misleads us. It is a fascinating topic. But to be honest, I think I am probably knit-picking. Conjurers, smoke and mirrors, and peculiar viewing angles may well be informative and amusing, but could there be more fundamental ways in which reality is masked by our senses? And where might we expect to find clues that it is so?

In order to answer this we might do well to remember that we humans have lived on this small planet Earth for a long time, and we are well suited to its environment. Consequently then, within our familiar world, we might expect our brain to be well able to read the signals and tell us what we need to know to live and survive here.

So let us move to another less familiar realm, where nature may not have equipped us read the world quite so comfortably. Though I'm not suggesting we move our location to some other exotic world. Rather, let's adjust our scale. It is in the macroscopic and the microscopic realms that we might be expected to find indications that our world isn't quite what our brains tell us it is.

**Spacetime**

Albert Einstein, a young man in the early years of the 20th century, was intrigued by a rather puzzling phenomena. The phenomena being that the velocity of light remains the same to any observer no matter where they are and what direction they are moving in.

Let's just think about what that means. Imagine you have a device; this device has a handle, a hoop and a digital display. By holding the device so that light travels through the hoop, let's imagine that the display can read the speed of the light as it passes through.

Now let's put you in space, wearing a space suit to keep you alive, and we'll throw in a super-fast space scooter for you to ride around on as well. As fast as your space scooter is however, let's imagine for now that you are stationary and you hold your light speed measuring device up to the sun so that you can read the speed of light being emitted. You will see that the speed of light is 299,792 kilometers per second.

Now let's hit the throttle of the space scooter and start moving towards the sun, and once we have picked up a decent amount of speed, say 100,000 kilometers per second, pretty fast, let's switch on the cruise control (actually, in space, cruise control simply means switching off the thrusters) and now you can use your device again and see what speed light is passing through it this time.

Common sense would tell us that the speed of light passing through the hoop has got faster.

That makes sense. You are travelling towards the sun and the light is travelling towards you so surely the light will be passing through the hoop at a faster rate. In this instance the speed of the light going through your device ought to be the speed of the light coming towards you from the sun, plus the speed you are moving towards the sun: 399,792 kilometers per hour, right?

Well no, it turns out that that is wrong. The light is passing through the hoop at the same speed as it was when you were stationary, 299,792 kilometers per second! So what is going on?

Well there are better qualified people than me to explain the intricacies of Einstein's theory of special relativity (not to mention general relativity which adds even more complexity), but let it be said that such was Einstein's genius he was clever enough to make sense of it. One of his (many) insights was that for light speed to remain constant to all observers then both time and space must vary depending on those observers' speed relative to one another.

Or to put it another way we might say that space and time are essentially the same thing and the degree you are passing through each of them depends upon your velocity.

This has all been demonstrated by the way. For example, very precise clocks which have been taken into space and left in Earth orbit for a while show that they have registered less time than clocks left here on the Earth's surface.

> *...the universe, whilst probably finite, has no edge.*

But it is at the cosmic scale that Einstein's revelations really throws our sense of understanding. It transpires, for instance, that the universe, whilst probably finite, has no edge. Does that confuse you? It certainly confuses me. My brain can literally not compute it.

Another thought before we move on. A particle of light is called a photon and therefore photons travel at the speed of light. Photons have no mass (they weigh nothing) so it might seem a silly idea if I ask you to imagine one wearing a wrist-watch, but please do your best and humor me. Now let's say that our photon is one of the many coming from the stars that light up our night sky, and have travelled many thousands of light years. And so, by our reckoning, have been on their journey for the same thousands of years.

But if our watch wearing photon was to check its watch on arrival, how long would it find it had taken to travel those countless miles? Relativity tells us that the answer is no time at all, the watch would not have moved a single second. So what, looking at things from our photon's viewpoint, are we to make of the vastness of space and time when it can be crossed in an instant?

The answer is nothing at all. As far as the photon is concerned space, as well as time, has shrunk to nothing. From this perspective, traveling at light speed you aren't travelling at all, you have to slow down for time and space to seem real.

## Quantum

Light is a rather confusing thing altogether, and the pioneering scientists of the early twentieth century certainly found it very intriguing. Not only did its speed, constant to any observer, have them scratching their heads, another fascinating question came when they tried to sort out what light actually was.

This was because, although they had established that light was composed of tiny particles, which they named photons, in some respects light also behaved rather like a wave, much like a wave which moves over the surface of a lake or sea.

To help resolve the question of whether light is, in fact, a particle or a wave various experiments and measurements were made (including the (to quantum science) classic 'double slit' experiment, I won't go into details here but I would recommend you look up and read about it (there's an entertaining YouTube link if you search for 'Double Slit Experiment explained! by Jim Al-Khalili'), ideas were put forward, tested, some discounted, some modified, and occasionally, accepted. Bit by bit they meticulously put it all together and eventually they reached the following, rather odd, conclusion, which I am somewhat simplifying here:

Photons are indeed particles, but the way they move around is nothing like the way normal objects that we are familiar with move around, rather they move around in accordance with something called a 'wave-function'. Now the wave function doesn't tell you where precisely the photon is at any given time, instead it tells you what the probability is that the photon will be at a particular place at any given time.

And it seems it could be anywhere! It's just that it is more likely to be in some places than it is in other places. And if you look at the areas of high and low probability, then you see the wave like pattern.

If this is not bewildering enough, it also turns out that the photon (or if it comes to that, any sub-atomic particle) doesn't really have any actual position in space or time, it exists in all its possible locations until it is 'forced', seemingly by observation, to reveal itself.

And if you don't understand any of this, or if you do but have difficulty accepting it, be assured you are in good company. As Niels Bohr, an early pioneer of quantum theory, once put it, "Anyone who is not shocked by quantum theory has not understood it."

*And if you don't understand any of this, or if you do but have difficulty accepting it, be assured you are in good company.*

### So what?

So what does all this talk of relativity and quantum have to do with the virtual reality world inside your head?

Well, it seems to me that if you were as big as the cosmos, or as small as a photon, then in order to be functional, your virtual reality system would need to be very different. And in particular your notions of space and time would need to be completely re-written.

It is a fundamental notion of Yuwayol that all things are merely different aspects of the same thing, that the universe is not, as we see it, a big open space with lots of things in it, drifting through time. Rather, it is just that our brain has evolved, in accordance with our local environment, to show us it that way. Like the color blue, or the note C-sharp, both time and space do not exist 'out there' in quite the way we see them.

*Like the color blue, or the note C-sharp, both time and space do not exist 'out there' in quite the way we see them.*

Then how exactly should we perceive the world we live in if we are to make sense of it?

Actually science has been wrestling with this problem for some time now, and the feeling has grown that we need a new perspective in order to make sense of what we know. As I have said before, I am no scientist. However, I am tempted to wonder if the Yuwayol perspective might have something to offer here.

# Chapter 6

# The Yuwayol Perspective

*"True self is non-self, the awareness that the self is made only of non-self elements. There's no separation between self and other, and everything is interconnected. Once you are aware of that you are no longer caught in the idea that you are a separate entity."*
**Thích Nhât Hanh**

*"It turns out to be very difficult to devise a theory to describe the universe all in one go. Instead, we break the problem up into bits and invent a number of partial theories. Each of these partial theories describes and predicts a certain limited class of observations, neglecting the effects of other quantities, or representing them by simple sets of numbers. It may be that this approach is completely wrong. If everything in the universe depends on everything else in a fundamental way, it might be impossible to get close to a full solution by investigating parts of the problem in isolation. Nevertheless, it is certainly the way that we have made progress in the past."*
**Stephen Hawking**

## Two big questions

As alluded to in this book's introduction, the rational for Yuwayol is best understood in the context of two questions which have kept philosophers

*Now is the time to begin with that question of why there is such a thing as anything.*

busy for millennia. That is: "Why does anything exist?" and "What am I?"

Now before we go any further, let me emphasize: Yuwayol does not provide us with an answer to either of these questions. Well, not quite. But if Yuwayol is correct, then for reasons that will be explained, I think it takes us pretty much as close as we can get.

In the chapter "*The Misleading Nature of Perception*" I have dealt with the notion that the way we experience the universe around us can at best be thought of as a mere representation, and at times a misleading one at that, of what is really out there. Prior to this, the chapter "*Levels of Understanding*" discussed how it is difficult for us to even know how close we are to understanding our universe.

A word of warning however, the following section contains some mathematics. It's not terribly advanced mathematics (sorry if you've heard that before and been taken for a ride), but I know that even the simplest math will send otherwise rational people running for cover.

So, if you wish, you may skip the next bit; there will be a recap following which will allow you to stay with the discussion. But really, you might as well give the next section a go. It really isn't so hard (honest!).

## The power of zero

Let me start by making clear that I don't really think we can ever really know, in a way that will satisfy us, why something exists rather than nothing at all. But we do know that something does exist, so maybe, reluctantly (very reluctantly, if you are like me), we just have to accept it.

But if we do have to accept it then it seems to me that we also have to accept that what does exist is most likely there because it has to be there. If that feels a little abstract perhaps I can put it another way – it would seem that something exists because, for whatever reason, you just can't have "Nothing".

So, if you can't have Nothing, then the question is – why not? Well, perhaps mathematics might have an answer for us. If physics has taught us anything then it is that the universe is a deeply mathematical place.

Now before we go any further, some of you might want to raise an objection at this point. Namely that, if, by nothing, we really mean nothing (which we do) then does not that even include mathematics? So therefore we cannot invoke mathematics before we understand why there is such a thing as mathematics at all. Well the point is valid, if arguable.

But I don't intend to argue. Rather, I would like you to just bear with me. I do not intend to provide a mathematical explanation for how something comes from nothing. Rather, as I am proceeding on the basis that whilst we may never know

*... while it may be true that mathematics cannot give us a definitive answer, being the most elemental of the sciences, maybe it can give us a clue.*

the answer to why anything exists, we have to assume an answer certainly does exist out there (because we know there is something). And while it may be true that mathematics cannot give us a definitive answer, being the most elemental of the sciences, maybe it can give us a *clue*.

So, let's assume that nothing can be represented by zero (I admit that this is a contestable assumption, however it is the best we can do), let us see if we can get a non-zero answer just from zero.

Well zero plus zero doesn't help, because that still leaves us with zero. The same applies to zero minus zero as well as zero multiplied by zero.

What about zero divided by zero?

Now this is more interesting. Keying zero divided by zero into my calculator gives me 'error', this is because it seems that computers don't like dividing anything by zero, but if we try to visualize intuitively zero divided by zero I feel there are three candidate answers: zero (because if you try to share nothing between no people then no one gets anything), one (because any number divided by itself equals one), and infinity (because you can multiply zero by an infinite amount and you will still get zero (actually you can use this argument to make zero divided by zero equal to any value)).

Well that's two out of three non-zero answers for zero divided by zero. That's not bad is it?

My favourite candidate though is zero to the power of zero. For those of you who haven't done any math for a long time, a quick recap. Take the following bit of arithmetic as an example:

$$2x2x2$$

Here, the number two appears three times in a multiplication. Mathematicians describe this as two to the power of three (which if you have worked it out equals eight). In shorthand this is written as $2^3$. Similarly:

$$3x3x3x3$$

This is the number three appearing four times in a multiplication. So we can say that this is three to the power of four (which equals 81), or $3^4$.

Now one of the reasons for dealing with this kind of sum is that it is helpful in dealing with dimensions. When we look at the physical world around us we perceive three dimensions; we can describe these dimensions as length, width and height. If you imagine a tidy pile of bricks five bricks long, and five bricks wide, and five bricks high, you will have (using our shorthand) $5^3$ bricks, or, using longhand, 5x5x5 bricks, which is 125 bricks in the pile.

So what if we were to use this notation in order to describe nothing? Let's say we have nothing, how can we measure it? Well it has no size, because it is nothing, and it has no dimensions, because it is nothing. So we have zero to the power of zero, or in shorthand: $0^0$ (In longhand this would just have to be a blank space!).

*... Zero to the power of zero equals one!*

Now, curiously, $0^0$ does not equal, as you would be forgiven for thinking, zero. Zero to the power of zero equals one! If you have the function on your calculator (the keys to press on my smart-phone are [0], [yx], [0], [=]), you can try it yourself. Most calculators will tell you that zero to the power of zero equals one.

Now I won't explain why this curious result comes about, if you feel the need to know then please research it yourself, but does it not strike you as curious that a fairly simple sum involving nothing but zeros should result in a non-zero answer? The clue we were looking for perhaps?

Now despite what it says on your calculator it is only fair to mention that most mathematicians would probably prefer to define zero to the power of zero not as one, but as 'undefined'. The notion of $0^0$ in this context, they would quite likely point out, is nonsensical. However I am not sure we should dismiss the idea that the universe might indeed be not just undefined but quite likely undefinable, and its existence, a product of nothing, quite nonsensical to our ways of thinking.

## One what?

For all the math-phobes who decided to miss out the last section let us briefly recap.

We wondered if the curious fact that zero to the power of zero (zero multiplied by itself zero times) equals one, might give us a clue as to why it is that something, rather than nothing, exists. It is important to note here that I do not claim that this is our sought after answer. I am not altogether sure, despite my reasoning above, that it is reasonable to describe nothing as $0^0$. Rather I am suggesting that this sum might be giving us a hint, if not of how something comes from nothing, then of something very important about the nature of something that comes from nothing – its oneness.

I intend to provide more reasoning later concerning the oneness of things, but for now let us presume that the thing that emerges from nothing should be regarded as a single unit. Forget the math, instead let us suppose that, for some reason we may never really be able to conceive, Nothing is untenable. It would be inevitable then, that instead of Nothing, we have to have some thing, or to be precise – One Thing.

So what is this thing? What exactly is this One Thing that simply has to be? Well the simple answer is that the one thing we are talking

> *...for some reason we may never really be able to conceive, Nothing is untenable.*

about is everything that exists. Does that sound like a contradiction? You may well be thinking that everything that exists is lots and lots of things?

Well that's the thing about Yuwayol. Yuwayol says that every thing is the same thing really.

And that includes you.

**Describing the totaliverse**

So, how can we say that every thing is the same thing? In order to help explain I'm going to introduce you to a new word which I have just made up – "totaliverse".

> *The definition of totaliverse is simply – everything.*

The definition of totaliverse is simply – everything. That is everything that exists anywhere. If there are other universes then it describes them as well as this universe. If there is a god or a Valhalla or an infinite number of giant tortoises then he/she/it/them are all included also.

And the totaliverse also includes all of these things in the past and in the future, from whenever time started, if it ever did, to when time ends, if it ever will. There simply is no such thing as something out side of the totaliverse because the totaliverse, by definition, does not have an outside.

In the chapter "*The Misleading Nature of Perception*" I touched upon how it is that the way we experience the world around us is really nothing more than a representation, or shall we say an interpretation, of that world. Colors, sounds, smells, tastes and touch are constructed for us by our brain so that we can live without the inconvenience of, for instance, walking into lamp-posts without knowing they are there. They also enable us to perform quite sophisticated tasks such as making a nice pot of tea.

The experiences our senses provide us with serve us well, and were honed by evolution to help our ancestors and us live, survive and breed in the environment we live in. This idea, that effectively we live within a world constructed for us by our own brain, is quite a peculiar one if you've never really considered it before, but I hope that you now understand that it is intrinsically true, and that features we experience such as color, though they may represent something out there, are illusory in nature.

Now, Yuwayol presumes that the three dimensional space we seem to live in, and time through which it seems to pass, is as much an illusion as color. It is a framework constructed by our brain's virtual reality system within which to place objects and events.

That isn't to say that space and time don't exist, it's just that the way we experience time and space may in some respects hinder us from understanding their true nature.

The chapter "*The Totaliverse and Yuwayol*" speculates on the phenomena that space and time actually do represent, but for now let's just consider one of Yuwayol's central ideas: The totaliverse should not be regarded as a great empty space passing through time with lots of things (things like stars, planets, teapots and you and me) in it. This is merely the way we have been hard-wired (as we have adapted to our environment) to see the things.

Rather the totaliverse should, at its most fundamental, be seen as a single entity, where all the things that we perceive, including ourselves and each other, throughout the different times we perceive them, are of the many (possibly infinite) *aspects* of it.

**The totaliverse simply cannot be defined...**

It would be easier if I could now present you with a picture of the totaliverse that you could hold in your mind; a definitive description of the totaliverse. But the point here is that there is no such description, nor can there be.

The totaliverse simply cannot be defined for the very simple reason that it isn't like anything else (because it is all that there is). It is the only one and exists because Nothing cannot. There is no mirror that can be held up, in our imagination or anywhere else, which we can use to reflect the totaliverse. It is at once all things, the one thing, and (much like the way some mathematicians prefer to label $0^0$) it is undefined (and therefore undefinable).

(And because the total sum of positive and negative energy, which is what the totaliverse appears to be made of, may actually be zero, it might well be thought of as not being there at all!)

So we are left with only being able to describe a single aspect of the totaliverse at a time. Which is pretty much what our brain does, and we experience each of these aspects as *things*.

Now, to give a notion of what I mean by an aspect of the universe, I hope that it will help if I provide you with a couple of Yuwayol-totaliverse-similes which might help you to understand the concept. But please bear in mind that none of these totaliverse-similes are perfect; the totaliverse, as we have said, cannot be defined.

Nevertheless:

### Simile 1 – pack of cards
With Yuwayol it may be useful to think of the totaliverse as being like a pack of cards. Now a standard pack of cards contains (forget the jokers)

fifty-two individual cards. There are thirteen 'numbers' (ace, two, three... through to ten, Jack, Queen and King) and there are four 'suits' (hearts, clubs, spades and diamonds).

Now an aspect of a pack of cards can be thought of as being the cards of the pack stacked in a particular sequence. You can imagine shuffling the pack, and could do so millions of times, arriving at a different sequence each time (actually the number of sequences is 80,658,175,170,943,878,571,660,636,8 56,403,766,975,289,505,440,883,277,824,000,000,0 00,000 which is a very big number, so big that whenever you shuffle a pack of cards reasonably well it is highly unlikely that the sequence of cards in that pack has ever been duplicated, even if you go back to the time playing cards were first invented).

The overwhelming majority of these sequences are very chaotic and have little rhyme or pattern (for instance a sequence might start randomly with the 8 of clubs, 3 of spades, jack of diamonds, 3 of clubs, 5 of hearts...), whilst a much smaller number of sequences are much more organized (and so might start with the ace, two, three to the king of diamonds, followed by ace, two, three to the king of hearts, followed by the other two suits in the same numeric order).

Now if each of these many sequences represent an aspect of the pack-of-cards-totaliverse, we can see that they are each the entire totaliverse, but at the same time have their own distinct identity, or character. So

*The planet Jupiter, the BBC, an anthill, a teapot, you – these can all be thought of as a different shuffle of the totaliverse pack of cards.*

(Yuwayol says) is any thing in the totaliverse: The planet Jupiter, the BBC, an anthill, a teapot, you – these can all be thought of as a different shuffle of the totaliverse pack of cards.

## Simile 2 – wavy rope

Or we can think of the totaliverse as being like a length of skipping rope. Perhaps you have played, most likely as a child, with a long piece of skipping rope. It works best with two, each holding the rope at either end. Now by waving the rope from side to side, or up and down, you can get the rope to create wave patterns.

These patterns are neatest when they divide the length of rope into equal sections. In the diagram below we see a variety of wave lengths along our imaginary skipping rope. In the last wave pattern we see a wave pattern which is a combination of the first five simpler wave patterns.

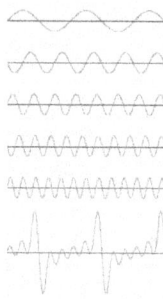

For Yuwayol we have to visualize for our skipping-rope-totaliverse a very long length of skipping rope with a really unimaginably huge number of possible wave patterns (actually the length of the rope is irrelevant but possibly easier to imagine a lot of wave patterns if we say it is long), and any thing in the totaliverse can be represented by a combination of these wave patterns (including you, though the wave pattern which would represent you would be very much more complex than this), but we must not forget that they are all played out on the same skipping rope, which in this simile represents the entire totaliverse.

There may be other descriptions which might act as similes for the Yuwayol totaliverse but these two, the pack of cards and the wavy rope, should be enough to be getting along with. The important point here is that, according to Yuwayol, in any ultimate understanding of the totaliverse (as speculated upon in the chapter "*Levels of Understanding*") all things within it can be regarded as individual aspects of it all.

**Your self is the totaliverse**

So on to the second of our two big questions – what am I?

Well it follows that if everything in the totaliverse is a single aspect of the totaliverse then I must be a single aspect of the totaliverse, and so must you.

But what about that part of ourselves which we have discussed, the bit that is the essential self? Let's not forget that in this book the word "self" should be taken to mean precisely this. We have said that this self can only be experienced individually, your experience of your self is different from my experience of mine. Indeed, for all that you know, my self might not be there at all. When we ask "what am I?" we are basically asking ourselves what this self actually *is*. Some might say that the self is an illusion, which of course begs the question – who or what is experiencing that illusion? The question doesn't go away.

*...the thing which experiences the self is the totaliverse itself...*

Yuwayol neatly answers this question by stating that the thing which experiences the self is the totaliverse itself, and it experiences it as you, through your thoughts and feelings. Just the same as it experiences it through me, and my thoughts and feelings. And indeed the same can be said of every other being which has the experience of self.

91

Looking at the pack of cards simile you are the pack of cards shuffled one way, and I am the same pack of cards shuffled a different way, and this accounts for our different looks, memories and personalities. But we are the same entity, you are me and I am you, just different aspects of the same thing. We share the same self and our self is the totaliverse.

In the chapter "*What am I? (The Question)*", we discussed the puzzle of how, despite knowing that we exist, the entity of our self eludes us however deeply we look into ourselves. I mentioned this might be because we are looking in the wrong direction. In fact our self is the product of the existence of the totaliverse. It is the totaliverse, in all its expansive undefinable glory, and that is why science cannot locate it in the brain.

Now, it may seem like hubris to claim to be the totaliverse, but if we put that to one side for a while (I'll deal with the point later on) and think about it, it does rather nicely answer the question – what am I? Furthermore it somewhat surprisingly manages to do so in a way which means that many of the philosophical problems we have previously discussed in the chapter "*What am I? (The question)*" just melt away.

> *...the philosophical problems we have previously discussed ... just melt away.*

92

For instance, remember the "Beam Me Up" thought experiment, where we considered the idea that you might be able to use the Star Trek transporter to have yourself beamed to some location and, instead of sending you, it malfunctioned and generated two copies of yourself (being a thought experiment we are not really interested in the fact that this is all highly unlikely)? So, if this did happen, which of the two copies would really be you?

Well, with Yuwayol, they both would (though it would be no less embarrassing a situation!). In fact this scenario gives us no more difficulty than acknowledging that your self is the same as my self. The two 'yous' would now be distinct (though presumably very similar) aspects of the same thing and your self would be both of them, so would my self come to that!

In a similar way Fred's dilemma is also solved. Fred would certainly have not have existed had his mother moved out to Australia when she had been a young girl. But Fred's self would have been any child that had been produced as a result of any conceiving done by his mother once she had grown up.

In addition, the problems we raised in our thought experiment involving a genetic change that we imagined had taken place at Fred's conceiving, are solved. Whether Fred or Fred2 had been produced at conception, Fred's self would still have been the result.

And what about the point made in the same chapter about us changing as we grow older? As we grow, at every stage from young babies to elderly men and women, our bodies change dramatically, our ideas change and our likes and dislikes also adjust over the years. So should we consider each stage of life to be that of different people? Well maybe in some respects yes. But according to Yuwayol, you've always been the same entity.

I repeat: your self is the totaliverse.

## Why you?

So when we asked earlier what it was that must exist if Nothing cannot, Yuwayol provides us with a simple, extraordinary, mind-blowing answer – you, or more specifically, your self.

I use the term mind-blowing because it really does require us to start thinking about our self from a very different perspective. For instance, one objection you might have to this idea is that it might strike you as unlikely that the totaliverse turns out to be you. Why you? You may well ask. Why not some other entity? But in fact this question comes from a mind-set which has got used to the idea that you share the totaliverse with lots of 'others'.

But in Yuwayol this is a misleading perception. In Yuwayol there is and only ever has been the self (not always in a conscious form). The self is 'it', and due to the oneness of the totaliverse there cannot be any other.

*The self is 'it', and due to the oneness of the totaliverse there cannot be any other.*

In fact Yuwayol merges our two big questions together such that one might just as well ask "What is everything?" rather than "What am I?", and "Why do I exist?" rather than "Why does anything exist?".

I can understand if you are of the opinion that, just for the sake of simplifying our two big questions, I have sacrificed quite a lot of what you may be accustomed to regarding as common sense! But such is the profundity of these questions, and our inability to so far answer them, should we be so surprised if our conclusions turn out to be both remarkable and unexpected?

So how does Yuwayol stand up when we consider what we know about the world we live in? Is this an idea that science should scorn? I am happy to concede that Yuwayol cannot be considered 'science'. At least not yet. However in the next chapter I hope to show that here is an idea that might potentially be a part of scientific thought yet to come.

# Chapter 7

# The Totaliverse and Yuwayol

*"If you wish to make an apple pie from scratch, you must first invent the universe."*
**Carl Sagan.**

## Disclaimer

In the opening chapter of this book I emphasized that Yuwayol cannot be described as scientific. I have no way, nor even can I suggest a way, of proving or disproving Yuwayol, which pretty much makes it a non-starter as far as the scientific method is concerned. So for that reason, for now it must remain in the domain of philosophical speculation.

However, I think that, based on our current level of understanding (see the chapter *"Levels of Understanding"*) there are ways in which we might speculate on how Yuwayol might fit in with our current knowledge of the cosmos. Some of these ideas may seem somewhat technical to some of you, whilst being rather crude and over simplistic to others. For this I apologize in advance, whichever category you might belong to.

*...Yuwayol is not as outlandish as you may be thinking, and fits in better with current knowledge than you might realise.*

Hopefully this shouldn't matter too much though, as my objective is not to provide some kind of authentic Yuwayol guide to cosmology; that would be beyond the scope of this book, and way beyond my means anyhow. Rather I am intending to show that Yuwayol is not as outlandish as you may be thinking, and fits in better with current knowledge than you might realize.

As I also explained in "*Introducing Yuwayol*" there is nothing supernatural about Yuwayol. (In fact the reverse is true, Yuwayol implicitly rules out any kind of ghostly supernatural or spiritual otherworld because Yuwayol says that the totaliverse is all that there is and it is all one, there cannot be an 'other'.)

If Yuwayol should at any point become provable, then we should expect that proof to be provided by mathematics and science, and, by the way, if Yuwayol was found to be contradictory to anything that has been mathematically or scientifically proven, then the inescapable conclusion is that it would simply be wrong.

Perhaps we should regard Yuwayol currently as a speculative top-down exercise in trying to understanding the totaliverse, whereas science provides a more reliable bottom-up methodology. Here, I am trying to signpost routes by which a future meeting between these two approaches might take place.

I should add that many of the ideas here are not mine, they are out there already being considered by greater intellects than my own.

So, in no particular order, and in a brainstorming kind of way where we are just throwing a few ideas around, what follows are some thoughts on the topic. Once again, they are quite speculative so please don't get too hung up on them.

### If newton had been right yuwayol would be wrong

A point worth mentioning is that if we still held to a Newtonian view of the cosmos then it would be hard to see any case for Yuwayol.

Isaac Newton may well have been the most influential scientist in history, in particular his work in optics, calculus, motion and gravitation was both brilliant and ground breaking. But at least as important as this is that his work established that the workings of nature might be defined by a set of, largely mathematical, rules, and that the motion of the stars in the heavens were just as subject to these rules as was the motion of objects (such as apples falling from trees) here on earth. He was largely responsible for setting up a conceptual model of the universe which was to hold sway for several centuries.

*Newton's vision of the universe was basically one which fitted in with what we might think of as a common sense view.*

Newton's vision of the universe was basically one which fitted in with what we might think of as a common sense view. What I mean by this is that his view of the universe fits fairly comfortably alongside the way our brains are programmed to see the world around us.

In this view our world is something which exists in a three dimensional space which moves at a regular rate through time. This otherwise empty space is full of objects, such as planets, stars, apples, apple pies and teapots, not to mention you and me. And these objects exert a gravitational pull depending on their mass upon each other, so that a very large object (such as a star) exerts a far greater gravitational pull than a small object (such as an apple).

So all these objects move around in accordance with Newton's famous laws of motion, their interactions being limited to those occasions when they happened to bump into one another or at least get caught in one another's gravitational pull. (As well as gravity magnetism was also known about in Newton's time but let's not get overly detailed.)

In this view of the cosmos time and space were a given (meaning they were simply there), probably thought to be infinite in extent, and the framework within which the separate objects did whatever they did according to Newton's laws.

If this view of the universe still held sway then I think it would have been difficult to conceive (at least for me) of Yuwayol. Yuwayol is based on the idea that all things are the same thing, not that they are independent entities.

Almost three hundred years later the genius of Albert Einstein punctured Newton's vision of the cosmos when he showed that time and space are the same. Nowadays cosmologists talk about the 'fabric' of space-time, or the space-time continuum. In other words space is seen as having at least a form of substance rather than being the empty space we perceive. And whilst Einstein himself failed to predict it (something he once referred to as his 'greatest blunder', he introduced a 'cosmological constant' into his equations to actually prevent the prediction) Einstein's equations pointed to an expanding universe, a phenomenon later proven by the observations of Edwin Hubble.

Space is getting bigger, it used to be smaller, and going back in time to the Big Bang it was tiny!

Now I would be heavily remiss to suggest that Einstein proves Yuwayol. However there are good grounds for stating that our understanding of what space and time is has become such that we can no longer be sure of their real nature. Issac Newton's simple three dimensional space has gone, and we can state with some confidence that space and time are not as we tend to perceive them to be.

*...we can state with some confidence that space and time are not as we tend to perceive them...*

For Yuwayol to make sense then space-time can only be representative of some *thing*. And any *thing* should be thought of as an aspect of the totaliverse as a whole. If we were to borrow our wavy rope simile of the totaliverse from the previous chapter then space-time would have its own wave pattern. I'm guessing quite a basic one.

## Wave-particle duality

When people first come up against the principles of the quantum world they commonly come up against a problem – it doesn't seem to make sense!

One particularly troublesome concept is that of wave-particle duality.

Over the past century or so we have discovered various elementary particles. By elementary particles we are referring to what as far as we can tell are the basic building blocks of everything we know that exists. An atom is not an elementary particle. It is made up of electrons, which we believe to be an elementary particle, and protons and neutrons, which are not. Protons and neutrons are made up of quarks. There are various types of quark but the ones making up protons and neutrons are called up-quarks and down-quarks. Quarks are also, so far as we know, elementary particles.

There are other particles that have been discovered, not to mention plenty more which have been hypothesized but have not, so far, been discovered. We need not discuss them all here.

Now one of the many puzzling features of particles and atoms is that whilst they quite definitely act in many respects as you would expect a particle to behave, they have very definite wave like qualities. In a previous chapter, *"The misleading nature of perception"* I mentioned the famous 'double slit' experiment which highlights this seeming paradox very neatly.

Since our common sense tells us that a particle and a wave are two very different things we find ourselves scratching our heads trying to understand what is actually going on here.

The conclusion is that we can think of sub-atomic particles as particles, but that they move around in such a way that they are more likely to be in certain places than they are in another, and the likelihood of each possible destination can be calculated. And when you look at the pattern in space created by all possible destinations we see a wave pattern. This is very different from the trajectory of larger familiar objects, such as a ball rolling along a pool table, which moves in a well-defined line.

It is understandable to wonder how this can be so. In the double slit experiment how can an atom 'know' that there is two possible slits to go through and act accordingly? Actually the question goes even deeper than this. Remarkably, years of experiment and calculation has shown that quantum theory necessitates that the wave function for any particle can only be what it is if it 'knows' all possible locations in the entire universe (I use the word 'knows' advisedly, no one is suggesting that particles really know anything at all). Which isn't bad for a tiny particle!

It simply seems to defy the way we think things should work. Many quantum physicists respond by pointing out that nature is under no obligation to behave the way we think it should, which is a good point. However, to behave

> *It simply seems to defy the way we think things should work.*

as consistently as it does, surely nature should behave in a way which makes sense at some level? Scientists have good reason to avoid philosophy and a lot of them do, their concern is to prove what works and what doesn't.

However, I think that if the quantum world is ever to make any kind of sense to us it will be by understanding how our view of the universe is profoundly misleading us (at least as far as the quantum realm is concerned).

So could Yuwayol help us here? I would say potentially yes.

In our mind's eye we imagine both particles and waves to be things which operate in the space around us and in time. However Yuwayol is based on the notion that neither space nor time are as we perceive them in our mind's eye. Instead we should take an elementary particle and treat it, like all other things, as a one aspect of the totaliverse.

Now when we think of it this way we have, I think, a foundation for being able to see a way in which we can, in one sense, perceive this particle as being a particle, whilst at the same time, and in another sense, (because space is illusory) understanding that its influence spreads out into the entire universe in a wave like manner. Just a thought.

## Block totaliverse

> *...your future...*
> *is just as fixed as*
> *your past.*

In our common everyday experience, time seems to flow from one moment to the next. Yuwayol contends that this is illusory. It says that, whilst time does exists (and what it is will be speculated on later) the flowing of time, as we experience it, is not real. This means that your future (or futures, according to some many-world theories), from a cosmological perspective, is just as fixed as your past.

In other words, the past, the present and the future can be thought of as all existing at once (this clearly has implications for our notions of free-will, something we discuss in the chapter "*Getting your head around Yuwayol*").

One way to visualize this is to picture the three dimensions of the space around you and then add time in as a new dimension. So if you could see yourself in this way (which admittedly isn't easy to visualize because it means thinking in four dimensions rather than the usual three) then your body would look like a long creature snaking through the dimension of time. At one end your body would be shaped like a new born baby, and at the other it would be (presuming you will live to a good age) an old person about to die.

This picture of fixed time is sometimes referred to as a 'block universe'. Now if we were to add in any other universes that may or may not be out there and put them into the picture, including their time lines also, we get a block totaliverse.

Now we should note that this is simply one way of picturing the totaliverse which is not too dissimilar from the way we normally picture it, only we have added a new, non-moving, dimension which is time. This gives us a view of the totaliverse which is static, like a four dimensional block, hence the name.

This view of the totaliverse is not altogether complete, in a similar way that the skipping-rope-totaliverse or the pack-of-cards-totaliverse are also incomplete. The block-totaliverse view (remember that the totaliverse may not be definable) simply provides a way of seeing the totaliverse which may be helpful.

### Not a process, a pattern

One objection to the block universe is that it may seem to deny the existence of any kind on progression taking place. Once again from our human perspective life is viewed as a series of events which seem to progress in a fairly orderly way. If you are thirsty and you drink some water then you are not so thirsty anymore; if you drop a teapot onto a hard floor it will smash into pieces; when the sun rises each morning the darkness outdoors is replaced with light. So events do seem to flow along in accordance with a set of rules which, at least in everyday terms, are relatively predictable.

In a block universe these rules are no longer to be thought of as applying to a series of events, but rather as applying to a kind of fixed pattern which is configured in accordance with those rules. A fixed pattern, it should be noted, can follow a set of rules in much the same way as a sequence of events do.

So, if we are not any longer saying that time flows, then why is it that we humans experience time the way we do – as a flowing entity with a past, present, and a yet to be determined future? To answer that we need to look at a phenomenon known as entropy.

### Entropy in time

Entropy is quite a complex thing to describe. But for the sake of our discussion we can be simplistic and say that entropy can be regarded as a measure of disorder within a system.

So as an illustration imagine a water tank (which we can think of as a system) with a divider in the middle. On one side of the divider you have clear water, and other side you have water colored with a red dye. You should consider this as an ordered state in that the red dyed water is neatly separated from the clear water. Now what if you were to remove the divider? In the first moment after the divider comes out the clear and red water is still separate, and therefore still well ordered. But very quickly the clear and red water begin to merge, increasing the disorder, or to put it another way, the entropy of the water tank increases.

We can imagine that quite quickly the clear water and red water has completely merged and, in this respect, the entropy has reached its peak, where it remains.

*...in any system entropy increases, or remains constant in time...*

The important thing to note about this example is that entropy increased with time. This, basically, is known as the second law of thermodynamics. It is an observable law of nature that in any system entropy increases, or remains constant, in time. Even if you were to sit and watch the tank of water from now until the end of the world you would be very unlikely to see the clear and red water return to their ordered, low entropy state.

Now it seems that the second law of thermodynamics is surprisingly the only law in science that recognizes that time goes in one direction and not the other. For instance, Einstein's laws of relativity work perfectly as well in reverse time as they do in forward time. As do the laws of quantum physics.

As we have observed, for us humans, time does seem to flow. But let us think about it. At any moment we are only directly aware of the present moment, the future is undetermined, and we are aware of the past only because of our memories.

And this, I think, is the key to understanding why time does seem to flow, because the process of creating our memories requires an increase in entropy. (This might seem counter intuitive. You could be forgiven for thinking the forming of memories is increasing order rather than decreasing it. But memory formation requires that our bodies must burn fuel to produce the energy for the brain to generate those memories, and the process of fuel burning, whilst it may produce some order locally in our brain, produces a net increase in disorder overall.)

So if an increase in entropy is required to form memories, and entropy only increases with time, it should be no surprise to realize that only the past can be remembered! And

*...this is why for us time flows in one direction only...*

this is why for us time flows in one direction only – from the past to the present and then on to the future.

## The problem with entropy and the big bang

You may have heard of cosmic microwave background radiation (CMBR). It is a constant radiation that comes from all parts of the sky and it is something that cosmologists are very keen to study and every so often it is talked about in scientific news stories. It was discovered in 1964 and this was significant because CMBR is something that had been predicted by the theory of the Big Bang and was seen by many as a clinching confirmation of that theory.

> *... at the moment of the Big Bang the universe was a singularity and ... must have had zero entropy.*

Since then the Big Bang has been accepted as the first moment of our known universe. There are, however, a number of issues around this event which have caused debate. One such issue arises because, at the moment of the Big Bang the universe was a singularity and, it is thought, must have had zero entropy.

In other words, at the moment of the universe's inception, there was no disorder. Now if we are to accept that the Big Bang was the moment that the universe sprang into existence, then how come it just so happened to do so with such a low state of entropy?

In order to address this problem let's refer back to one way of alluding to the totaliverse we discussed in the previous chapter "*The Yuwayol Perspective*". That is the pack-of-cards-totaliverse.

To recap, the pack-of-cards-totaliverse said that the totaliverse as a whole was to be likened to a pack of cards and all the 'things' in the totaliverse should be thought of as a different shuffle of this pack of cards. Each shuffle would produce a different sequence of cards if you were to deal them out.

Now in our pack-of-cards-totaliverse we can say that there is every possible sequence and each sequence represents a different 'thing' in our totaliverse. (You may remember that I told you that we have 80,658,175,170,943,878,571,660,636,856,403,766,975,289,505,440,883,277,824,000,000,000,000 things in our pack-of-cards-totaliverse). The overwhelming number of times you shuffle a pack of cards you get a very disorderly pack, that is to say that if you were to inspect the pack you would find that the cards are sequenced in a very random and disorganized order.

So most of the shuffles can be said to have a high entropy.

But there exists, amongst all these shuffled packs, certain sequences which do contain some order. There will be shuffles where, for instance, you might get all the cards in the suit of hearts perfectly sequenced (ace, two, three, four...etc.).

These shuffles have a lower entropy than the average shuffle (they are more ordered). So if there are any observers in our pack-of-cards-totaliverse (I think this is unlikely – you would need a lot more cards I think, but never mind), and if the second law of thermodynamics is operating, then there would be a tendency to perceive events defined by the lower entropy shuffles as being in the past.

Now there are a very few shuffles which can be seen as perfectly ordered. These shuffles would be like the pack of cards you take out of the packet for the first time, all the suits are separated and are in numerical sequence. In these packs there is no disorder. These shuffles then have no entropy, just like the Big Bang has no entropy in the real totaliverse.

Here then may lie the answer to the problem. Instead of thinking that the Big Bang just popped into existence from nothing, Yuwayol says that everything in the totaliverse exists as one thing, and this one thing exists because Nothing is impossible. And all the aspects of the one thing (the totaliverse) includes a thing which has no entropy.

And because an increase of entropy is synonymous with the progression of time it is reasonable to suppose that all things in the totaliverse will perceive the no-entropy event, the Big Bang, as being at the beginning of time (at least those things, like us, which have the ability to perceive and are aware of the Big Bang).

> *... it is reasonable to suppose that all things in the totaliverse will perceive the no-entropy event, the Big Bang, as being at the beginning of time...*

## Entropy is space-time

Being even more speculative than I already have been, we may wonder if we now have a way of viewing the nature of space as well as time.

In the chapter *"The Misleading Nature of Perception"* the idea was put forward that both time and space were not quite what we perceive. This is an important notion in Yuwayol because the way we perceive the many things in our daily lives suggest that all these things are all separate from each other in space as they flow through time, but Yuwayol says that all things are in fact the same thing. So our perception of space and time must be false.

Cosmologists tell us that at the instant of the Big Bang there was no time, nor was there any space. The other thing that we can say about the Big Bang is that there was no entropy, or disorder. Or at the least that there was very little time, space or entropy.

This is not an easy concept. The usual default picture that many people initially have of the Big Bang is that of a universe expanding out into space, but this is a misconception. The fabric of space, and time, are both created at the Big Bang and as tiny as the universe is at this point there is no concept of the universe having an 'outside' to expand into (another example of how the laws of nature defy our everyday notions of common sense).

It is thought that in the moments following the Big Bang the fledgling universe underwent a momentary huge surge of expansion (referred to as Inflation) following which the universe has been expanding ever since.

Recently it was discovered that not only does the universe continue to expand, but it is doing so at an accelerated rate (which on the face of it might suggest that the Big Crunch scenario mentioned in "*The Yuwayol Perspective*" is false).

It is important to understand here that when cosmologists talk about the expansion of the universe they are not really saying that the stars and galaxies contained therein are speeding away from each other, what they mean is that the space around all these objects is expanding. And it does so as the entropy in the universe continues to rise.

Can we then conclude that in order to understand the true nature of time and space, we might do well to regard them as being somehow a direct result of entropy? Perhaps with the quantity of space being influenced by the measure of entropy and time being influenced by a local measure of entropy increasing?

Once again, just a thought.

# Chapter 8

# Getting Your Head around Yuwayol

*"I'm the catcher in the rye*
*I'm the twinkle in her eye*
*I'm Jeff Goldblum in 'The Fly'*
*Well, who am I?"*
**Lyrics from Divine Comedy song "Gin Soaked**
**Boy".**

*"We are the cosmos made conscious and life is the*
*means by which the universe understands itself."*
**Brian Cox**

### A strange idea to get the hang of

In my opening chapter ("*Introducing Yuwayol*") I said that, whilst Yuwayol is a simple idea to explain, the ramifications may take a lifetime to get your head around. This chapter is intended to get you started.

### You and your self: clarifying yuwayol pronouns

Before going any further it is probably worthwhile revisiting the distinction I have made between pronouns such as "you", "me" and "us" etc. and my use of the word "self" as in "your self", "my self" and "our self". This is because understanding the difference will be ever more important as we progress from here.

I understand this can be confusing; it can sometimes confuse me! So in order to try and clarify this, let's specifically discuss how the use of "you" and "your self" mean different things in this book.

**You**

In Yuwayol the normal use of the word "you" refers to what it normally refers to in your usual everyday life. When you meet someone on the street and they say "Hi" to you, this is the "you" we are referring to.

*When you meet someone on the street and they say "Hi" to you, this is the "you" we are referring to.*

"You" have an identity, with a name and a physical appearance, your body, be it thin, fat, short or tall, healthy or not-so-healthy.

"You" have a personality, which makes you, serious, funny, interesting, dull, clever, wise or stupid, or a combination of all these things.

"You" also have an inner life which only you are directly aware of. This inner life comprises of many things. "You" have thoughts, feelings, hopes and desires.

"You" may have secrets.

"You" also have memories which give you access to your past.

All of these things have developed and evolved since the day you were born, and when you die, they will be gone.

"You" are temporary.

**Your self**

In the chapter "*What am I (The Question)?*" I encouraged you to think deeply about something which we described as the essential "you". The reason for this is because I wanted you to acknowledge the existence of an entity that, whilst it cannot be found or measured physically, it does seem to be there, apparently within you.

But in fact "your self", which is what we are calling this essential "you", shouldn't be thought of as something buried deep within. Quite the reverse. Yuwayol says that "your self" is something which equates to the entire totaliverse and that "you" (as defined above) are simply one aspect of it.

And so when we have talked about "you" being "me" then more correctly we should be saying "your self" is "you" as well as "me".

*...as Julius Ceasar, "your self" defeated the Gauls in 50 B.C. Clearly you didn't!*

The reverse is also true according to Yuwayol. "My self" is "me" as well as "you". In fact "your self", "my self", "our self", and all the other "selfs" we could use, essentially refer

119

to the same thing (the totaliverse), though in different contexts.

Which is why we can talk about how, being me your self sees the world as I do. Or how, as Julius Caesar, "your self" defeated the Gauls in 50 BC. Clearly you didn't!

So with that cleared up (I hope!), let's progress.

**Good news, bad news**

Shall we recap? As we stated in the opening chapter of this book, Yuwayol is the concept that all things are the same thing, and that includes you and everybody else.

Which means that your self *is* everybody and everything else. This has some pretty intriguing ramifications, of which some are good and some are not so good.

For instance, your self has walked on the moon (your self did so as Neil Armstrong and others). As various great composers your self has over the ages created our most beautiful and inspirational music. And your self, as various top international athletes, has also inspired millions by winning countless Olympic gold medals.

Sounds good? Well yes, I tend to think so. But there is a dark side to all this, your self has done some pretty terrible things too.

Your self was Adolf Hitler, and a large number of other unsavory characters as well (if you happen to be one of those curious people who thinks that Hitler was okay I won't waste time arguing, feel free to substitute Hitler for a villain of your own choosing as you read on).

Your self has also suffered at the hands of these people as their victims, as well as supported them, as their co-conspirators.

> *... as my tea pot your self is not alive and doesn't actually experience or care about anything.*

On a lighter and rather more flippant note, your self is also my teapot (which I value quite highly by the way)! However, because my tea pot is not a living thing (so far as I'm aware!), as my tea pot your self is not alive and doesn't actually experience or care about anything.

But it doesn't end there. Your self is everything, the whole of the totaliverse spread throughout the whole of space and time. But at one and the same time your self is almost powerless to prevent anything that happens within it. Your self is like some impotent and unknowing god.

Almost powerless did I say? There are times when you can change things. For instance, your self can change things right now, while your self is being you.

### I could never do that!

Sometimes we are witness to, or hear about, people who commit terrible crimes. I have already mentioned Adolf Hitler and you may have others in mind. For such people and such tragedies it is understandable that we invent terms such as "evil", as that would seem to set them apart, making them different from us.

I can easily understand that, when you contemplate people and crimes like these in the context of Yuwayol, you may recoil emotionally at the idea that your self might be implicated in such things, not least if you or your loved ones have been victim to such awful events.

And you may well feel that you could never do the things that Yuwayol claims your self (as well as my self I would remind you) has done. Please be assured that I am happy to believe you. For Yuwayol does not make such a claim. If we refer to the distinctions above to our use of the terms "you" and "your self" you should realize that *you are clearly not* Adolf Hitler!

However, what Yuwayol claims, which may be difficult for some (particularly if you believe you have a soul or are guided by your own unique spiritual element, which Yuwayol denies in any case) is that our self was indeed Adolf Hitler, and others too.

And our self, being born with his DNA, his upbringing, having all his experiences in life, being influenced by what he saw, heard and did, sharing his passions and loves and living his life in the time and places that he did, then our self would, and did, do the things that Adolf Hitler did.

It is an uncomfortable thought. But that is not you. You are still you. Yuwayol does not consider you to be somehow mysteriously tainted by any other life in this way. You don't (I personally hope) have those things in common with Adolf Hitler, and so there is no reason to suggest you would ever do what he did, or share responsibility for those things in any way.

> *Yuwayol does not consider you to be somehow mysteriously tainted by any other life in this way.*

## Yuwayol and the supernatural

Here I would like to re-emphasize a point made in the very first chapter about what Yuwayol is not.

That is - Yuwayol is not a supernatural or spiritual idea. You may be getting the idea of some sort of disembodied soul which flits between our physical bodies, experiencing our lives, like some form of simultaneous re-incarnation.

And I suppose you could kind of read it that way. But that isn't really what I'm suggesting. Rather what I'm suggesting is that your self (I could just as easily say my self) actually is the physical totaliverse, and every thing in that totaliverse is your self in a different guise. Whilst the idea might seem outlandish when you first encounter it, there is no suggestion here that some spirit, disconnected from nature, is able to possess physical bodies and live in them.

> *...if you think Yuwayol is a supernatural or mystical concept then you have utterly misunderstood it.*

In fact I will repeat the point that I have already made – if you think Yuwayol is a supernatural or mystical concept then you have utterly misunderstood it.

## So not supernatural, but is Yuwayol about a universal cosmic consciousness?

Actually Yuwayol, at its heart, has nothing to do with consciousness at all. It says that all living things are all aspects of the same thing, and it is on us as living things that the focus has been on for the purposes of this book.

But the logic of Yuwayol applies to non-living things also. Stars, buildings, T.V. sets and also, as I think I might have mentioned, my teapot! These things strike me as less important because they are not alive.

As my teapot your self doesn't really care about anything very much. The reason for this is that my teapot has no mind (as far as I'm aware!) and so is not conscious.

In fact, in all probability most things in the totaliverse are not conscious either. Nor, to my knowledge, is there any kind of central or universal guiding consciousness either. And even if there is, its existence is completely irrelevant to the Yuwayol philosophy.

It is important here to be aware of the distinction between the self we discussed in the chapter "*What am I (The Question)?*" and your consciousness. Undoubtedly consciousness is an important and precious thing to us, but so far as Yuwayol knows, it isn't what makes us exist nor is it what we are. The significance of consciousness is that it bestows self-awareness without which we could never have questioned what we are in the first place.

### Isn't Yuwayol a bit arrogant?

I can fully sympathize if you are thinking along the lines of "So, Yuwayol is claiming that I am the universe? It all sounds a bit self-centered doesn't it?" Well, yes at first glance it does seem rather arrogant, downright megalomanic was my own initial reaction. I've often felt dissuaded from Yuwayol for this very reason. Indeed I would probably have started taking the idea seriously many years earlier if it hadn't been for this.

Science has been knocking our perception of our status in the known universe for quite a while now.

Certainly there was a time when many humans believed that we were made in God's image and lived at the center of the universe, with the heavens orbiting around us.

Then Copernicus, Galileo and Keplar et al came along and showed us that the Earth, along with all the other planets in our solar system orbited around the sun, and so we lost our pivotal position in the solar system. If this was not bad enough astronomers, equipped with ever more sophisticated telescopes, probed further and realized that the sun was but nothing more than a fairly typical star, just one of a countless multitude in an unimaginably vast island of stars.

Could our status fall any lower than this? Yes it could. In the twentieth century we discovered that our island of stars, the Milky Way, our galaxy, was not alone. It was one of hundreds of billions of galaxies. We weren't just insignificant, in terms of size we as near as dammit didn't exist.

It didn't end with the vastness of space either. As I write this the NASA website tells me that the current age of the universe (from the big bang until now) is estimated to be around 13,770,000,000 years old (plus or minus 59,000,000 years!).

*...if some god-like observer was somehow able to locate our microscopic position in the vastness of space, if she blinked she might well miss us.*

Anatomically modern humans evolved around 200,000 years ago. This means that human men and women have existed in the universe for around a mere one sixty-nine thousandths (1/69,000) of its history. Even if some god-like observer was somehow able to locate our microscopic position in the vastness of space, if she blinked she might well miss us.

And that may not be the end of how insignificant we are. Some say that our universe is but one of many, possibly infinite, universes, making up what I have termed the totaliverse.

So, yes, for me to claim that I am the totaliverse would appear, to say the least, to be elevating my status somewhat!

And by asking you to consider that you are also the totaliverse, a different aspect of it, I can understand it if you have a reluctance to step in and go with the idea. The idea just feels quite literally too damned full of self-importance.

But hang on, I am not saying that I, Secret Phil, the human being who is writing this book, encompass all that the totaliverse is. I am saying that I am but one aspect of it. There are other aspects beyond count (you for one instance, and my teapot for another) that complete the picture of the totaliverse and in that sense my identity, as the simple, flawed, human that I am currently conscious of, maintains its insignificance against the backdrop of the vast totaliverse.

However, Yuwayol says that being an aspect of the totaliverse means that my self is the totaliverse, all aspects of it. But in truth this is not hubris because the same can be said for the 7,000,000,000+ other members of the human race who currently reside on our planet, not to mention all the other creatures who also live beside us, not to mention whatever other creatures may exist elsewhere in our universe, not to mention any other lifeforms who might inhabit other universes in the multiverse, not to mention all the other beings that have existed before us, or those who will come after.

It is all of our selves that are the same self.

## Of infinity

The pack-of-cards-totaliverse analogy which we have used in this book presents a totaliverse which is finite in respect to the number of aspects it can have. The number of distinct sequences we can order a standard pack of playing cards has a very large number - 80,658,175,170,943,878,571,660,636,856,403,766,975,289,505,440,883,277,824,000,000,000,000 to be precise. As big as that number is, though it is restricted to that very large number it is therefore limited in size. It is finite.

The skipping-rope-totaliverse analogy however is potentially infinite. That is to say the number of aspects that this totaliverse can have is unlimited, as we never specified a size for the skipping rope and, being imaginary, could potentially be unlimited in length. It therefore would have the potential to possess an infinite number of wave patterns.

*...we do not know whether the totaliverse is infinite or finite.*

Today, as a species we do not know whether the totaliverse is infinite or finite. In Yuwayol terms this means that we cannot know if the totaliverse has an unlimited number of aspects.

My suspicion is that the number is indeed infinite, but even if I'm wrong it is clear that the number of aspects the totaliverse has is beyond our imagining.

In order to illustrate the concept of infinity, the notion of a monkey randomly hitting keys on a typewriter for an infinite amount of time has often been cited. The idea is that given an infinite amount of time, this monkey will, eventually, randomly type the complete works of William Shakespeare.

Of course anyone observing, waiting for the monkey to actually do this, would have to be immensely patient. (One can imagine the near misses; I wonder how many times before the monkey achieved its task would the observer witness output along the lines of "Life ... is a tale Told by an idiot, full of sound and fury, signifying !@gibl$ix!es 43qt"...!")

Having an appreciation for the boundless nature of the infinite helps us to understand how, in its vastness, the totaliverse has managed to produce us. Like grains of sand on an immense beach, as remarkably complex as we are, we should keep it in our mind that we are just one tiny aspect in a huge array.

So, like the works of Shakespeare turning up in the monkey's endless manuscripts, it is perhaps not so surprising that something as extraordinary as we living things exist.

In fact even the mind boggling monkey-typist-metaphor doesn't really begin to do justice to the vastness of the infinite. We might well note that as well as Shakespeare's life's work, the monkey would also go on to produce literature which the bard could never even begin to dream of emulating. And the monkey would still have only just begun...

**Free will**

Each day we all have choices to make. We can decide to go out or stay in; in an election we can elect to vote for left wing party or we can opt for the right. We spend much of our lives trying to decide between various options and the outcome of these decisions will often have a profound effect on the way our lives unfold.

Yet the notions of the totaliverse as discussed in the chapter "*The Totaliverse and Yuwayol*" would seem to deny that any real choice takes place.

This is because Yuwayol says that our notions of time are mistaken. The past, future and present are all, in effect, present, and therefore set. So, your decision to go out is not a choice at all. You were always going to go out and therefore your so called decision is a mirage.

*...if we can't change the future then what's the point of doing anything?*

The idea that the future is all set out and fixed and that there is nothing anyone can do about it is not new. That such an image of the world is correct is often referred to as 'deterministic', and it makes us uncomfortable because it offends our notion of having a free will and this is an important part of how we see ourselves.

After all, if we can't change the future then what's the point of doing anything? There are two way in which we can address this issue:

The first is to question the validity of viewing it in this way in the first place.

The problem emerges I think when we first of all draw a conclusion based on a speculated higher level of understanding, which may be true (that the universe is deterministic), and then flip to a lower level of understanding and draw an inappropriate conclusion.

So, looking at a decision to do something from the perspective of the totaliverse, we might see that any decision you make was inevitable. But that does not mean that from your own human perspective you are not making choices.

You are, and from that perspective it is still beneficial to make those choices wisely.

The second way of addressing the problem of free will is to start by citing what we think free will actually means (rule 1: when tackling a question make sure you understand what the question is!).

I would suggest that free will means 'doing whatever you want within the bounds of physical possibility'.

You might think that 'doing what you want' sounds a bit glib in the face of such weighty matters, isn't free will about, for instance, higher moral decisions also? But I would argue that making a decision based on conscience or such weighty matters is still essentially doing what you want to do, because morally motivated people have reasons for wanting to do that.

As for what I mean by 'the bounds of physical possibility', I add that in so that we can exclude from consideration those things we would want to do but can't. I might want to fly to Neptune and back, the fact that I can't doesn't really have anything to do with what we are talking about.

*I would argue that in actual fact everyone always does what they want.*

Now I would argue that in actual fact everyone always does what they want.

It is true that things might not always feel that way. On a morning, when you need to get out of bed for some reason, you may feel uninclined to do so because you are still feeling drowsy. But you get out of bed anyway and on such occasion we may say that we are doing something even though we don't want to do it.

But on such occasions we are making a choice between two options, such as getting up to, let's say, earn some money at work, or staying in bed where we can catch up on our sleep. If we decide to go to work then our reasons for doing so has outweighed our reasons for staying where we are.

This may be because we don't want to lose our job, which would mean not earning the money or taking care of our family or so on. These considerations have triumphed in our minds over staying between our comfy sheets and eventually suffering the consequences.

And so, however miserable we might feel about it at the time, we are always doing what we want to do. So in that respect, yes, if we always do what we want then we absolutely do have free will.

The question which follows from this is: do we get to choose the things that we want? The ultimate answer to this I think is – no. We cannot, at any particular moment, change what we want to do at that moment.

That is not to say that we cannot decide to better ourselves so that we will try to want different things in the future ("the future" here could lie just a few seconds away). But even there we are making a decision now, based on what we want now. This decision may, if we are successful, mean that in the future we will be making decisions differently but at that time the decisions we make will be based on what we want then.

## So does life have any meaning?

Yuwayol simply states that all things are the same thing. And that includes all living things. There is nothing implied about there being any meaning in the world or that there is any kind of pre-ordained plan being played out. Nor does Yuwayol suggest that there is a god of any description in the background orchestrating everything that goes on.

However Yuwayol does not automatically rule out any of these things either.

For instance, God could be accommodated by Yuwayol but with certain caveats. Chief amongst these is that if god, or some god-like being, exists then he/she would be one thing in the multiverse (which by

> *... God could be accommodated by Yuwayol but with certain caveats.*

definition contains everything), and would therefore be one aspect of it. In this respect god would be just like the rest of us and so therefore would be your self and my self just the same as everyone else.

Frankly this seems to me to be a bit of a reduction in status when compared with how the main theistic religions (Judaism, Christianity and Islam) perceive god, and I am therefore inclined to say that the belief in god as taught by these religions does not sit easily beside the philosophy of Yuwayol. Though you may have your own ideas.

My own view is that meaning is something that living creatures bring to their own lives.

This might be through our relationships and interactions with those we know, perhaps in particular our families. Meaning can also come through the love of art and music and also through studying nature and the workings of the astonishing world around us.

Quite honestly, and speaking for myself, I find there is plenty of meaning in my own life, lots to be in awe of, so much to love and feel passionate about, so much of these things to be getting on with without having to invoke the supernatural, which in any case seems to me to trivialize rather than enhance my appreciation of the world I live in.

But that's me. You may see things differently.

# Chapter 9

# Moral Imperatives

*"Love your neighbor as yourself"*
**Jesus of Nazareth, Matthew 22:39, Bible (New International Version)**

### Moral compass

From what you have read in this book so far you may or may not be convinced by what you have read. But if you have come to the conclusion that there may be something to Yuwayol then it seems to me to be undeniable that a certain moral imperative does seem to follow on from that. Intriguingly this imperative

> *...it is in the interest of each of us individually to seek the interest of all of us.*

works by implying that it is in the interest of each of us individually to seek the interest of all of us.

In the opening chapter of this book I said that, for anyone who already possesses a healthy moral compass, then Yuwayol should make little difference to the way that compass guides them in their lives. But clearly that does rather depend on how we define morality in the first place.

This can be a very vexing subject as there are many contradictory and sometimes passionate ideas that people have of what makes for a sound moral life. But nevertheless, if you will permit me I will try and give you my own thoughts. This is not because, I beg you to note, I consider myself morally superior in any way (I can assure you I'm not, also I tend to regard self-righteousness to be the ugliest of human traits).

Rather I would like to examine this issue because I do actually think that Yuwayol could have the potential to motivate us to live better lives and I'd like to explain how.

## Being human

*We tend to care for our young, old and sick when we feel able.*

Humans are essentially social in nature. Throughout much of our history we have lived in small communities and this has nurtured certain sensitivities about how we ought to live together. We tend to care for our young, old and sick when we feel able thus providing a net benefit to all within the group.

Also, we are sensitive to notions of fair play, indeed it is striking how easily we are able to understand these notions even as young children.

Altruism, the ability to be concerned and act in accordance with the welfare of others, seems to be a universal human trait. Whilst the nature of altruism is hotly debated amongst those whose field of study is human behavior, certain key characteristics seem to stand out. The first is the way in which our altruism to others varies according to our relationship with them.

## Moral horizons

For instance, it seems clear that altruism is much more likely when people are dealing with their own kin. Normally this is the people they are related to, especially closest family members, and in particular children.

Indeed it is for kin that humans are generally prepared to make considerable sacrifices. It is very natural to be altruistic for close family, even to the point where in extreme situations quite normal people will sometimes willingly lay down their lives.

Many of us are also prepared to be altruistic towards those we consider to be our closest friends. This altruism is not normally as prevalent as it is in family ties but there are common exceptions. Indeed a closely bound group will often use words like "kinship", "brotherhood" or "sisterhood" in describing their relationship with one another.

Usually to a lesser degree than our friends our altruistic actions can extend to what we perceive to be our cultural group. What is meant here by "cultural group" differs from individual to individual. It might mean the village, or tribe they belong to; others might regard it as the country they are a citizen of; some, problematically in the modern world, might see their group as those whose members share the same religion, skin color, or ethnic origin as themselves.

Finally there is what might be thought of in a more universal scope, where we show kindness to people we may never meet. This sometimes comes to the fore when appeals are launched for the victims of disasters far away.

Of course the extent to which our generosity fades according to these categories, what we might think of as moral horizons, differ enormously from individual to individual. There are some who would not lift a finger for anyone, not even their own children, and there are others whose kindness to strangers seems quite remarkable. But I think these bands hold true generally for people as a whole.

> *There are some who would not lift a finger for anyone... and there are others whose kindness to strangers seems quite remarkable.*

## The deserving

Another important characteristic of human altruism is that we also discriminate when we feel that someone is deserving or undeserving of our kindness.

So for instance, there may be someone you know who has never hesitated to help you out when you have required aid, someone who you have always felt you can turn to when things are not good. So if, one day, this same person came to you and asked for your help, you would, because of their past kindness, feel more inclined to respond to their appeal.

Now many would be inclined to say that this is not true altruism because, basically, what is happening is a kind of trade off, a "you scratch my back and I'll scratch yours" kind of scenario which is of mutual benefit to both parties.

> *...many friendships do tend to proceed on a tit-for-tat basis.*

There probably is some truth in this. Most people probably feel ill inclined to give to a friend whom they know would not be prepared to give back should the circumstances be reversed, and many friendships do tend to proceed on a tit-for-tat basis.

But there are numerous examples where people give to others in need even where there seems to be little hope of the favor being returned, choosing to care for an elderly neighbor for instance. Such instances do seem to be the result of genuine compassion for people who, through no fault of their own, find themselves in need.

### Is altruism real?

However, the extent to which altruism in humans is genuine is a debatable thing.

Some argue that all, or nearly all, acts of kindness are motivated, consciously or otherwise, by a desire to maintain a positive reputation in the societies they live in. This in turn will incline others in that society to be good to them. So, not so much altruism, more quid-pro-quo.

I personally think this might be over-egging things a bit. Kindness feels like kindness to me, and I believe there is a joy to be got from actively giving of yourself which is a reward in its own right.

> *...there is a joy to be got from actively giving of yourself which is a reward in its own right.*

However it would seem that human altruism left to itself does have its limits, tempered by our moral horizons, beyond which our love for our fellows rarely travels. Controlled tests do seem to indicate that many of us, when it comes to doing the morally correct thing, are less inclined to do so if we think that nobody is watching!

**Following the rules**

As societies have expanded and merged from small close knit tribes to ever larger communities, which today can measure in the millions of individuals, steps have been taken to, if I can put it this way, supplement our natural altruistic natures. An obvious way that this has been done is by the implementation of various laws and rules.

> *The first issue with laws and rules lies with the problem of getting people to follow them.*

The first issue with laws and rules lies with the problem of getting people to follow them.

One way to deal with this is to try and draw up the laws to be as fair to people as possible. If you do this then people's sense of fair play are more likely to tend them towards respecting the laws and therefore more inclined support them.

Another measure that is usually taken is to impose penalties against those who are caught breaking the laws, such as fines, jail sentences and so on. Ideally, our instincts for fair play also come to the fore here in that the penalties ought to be seen as measured and appropriate.

So rules can take us a long way in getting people to live well alongside each other if they are implemented wisely. Indeed, a simple way of thinking of morality might be to view it as a set of rules we should follow in order to get us all to live good lives.

### Thou shalt not

But there are issues with this approach, one being that rules, however well intended, often lack the sophistication to cover all eventualities. The Ten Commandments, defined in book of Exodus in the old testament of the bible, have traditionally been held up in western culture as a basic set of laws, or rules, which can guide us to live morally dependable lives.

The sixth commandment says quite simply "Thou shalt not kill". Now, I think that most of us would agree that killing is wrong and so therefore "Thou shalt not kill" is a pretty good rule to follow.

But hang on. Have you ever used bleach in your kitchen? That's killing bacteria, millions of them. Ever chopped down a tree or pulled up a weed from the garden? Now killing a tree or a plant might not be so important, but it is killing nevertheless, thereby breaking the sixth commandment.

If that seems a bit pedantic what about animals. (If you're a vegetarian and avoid using animal products then I guess you can sit this next bit of the argument out.) If you eat meat or wear clothes made from animals (such as leather shoes) doesn't that break the commandment as well? Okay you haven't killed directly but you have required someone to do the killing for you so you can do these things, so surely that makes you compliant in the crime.

You might say that I'm misunderstanding the commandment, that it is understood that the commandment is only meant to mean "Thou shalt not kill a human."

Well I dare say that is true, but this example does show the dangers of not specifying the rules correctly and we can always wonder if the literal translation of the rule is what was intended all along.

But I'm wondering if even "Thou shalt not kill a human" might actually be a flawed rule in itself. I think we'd all agree that it should be applied almost all the time. But aren't there some extreme exceptions? What if you are trying to stop some psychopath from murdering your family and you are only in a position to stop him by lethal means? What if you are a soldier fighting for your people and your country?

As I say, these are exceptional circumstances and I hope you never find yourself in such a position, but doesn't this demonstrate that the rules are not perfect.

## Don't get caught!

And the respect for law, however well administered, is always likely to be limited. There will be always those who willfully disobey it, especially when they can get away with it. I'm willing to guess that most of us have broken at least some minor law at some time or another.

As the prison inmate Fletch from the BBC T.V. comedy 'Porridge' once remarked to his colleagues, "we're all here for the same reason - we got caught!"

In fact I think very few rules, however well thought out and detailed they are, can define precisely what is right or wrong in every given situation.

We give our children rules to follow, particularly when we take them into social situations they lack the sophistication to appreciate, such as

*Children cannot always be expected to fully understand why these rules exist…*

when we take them to a relative's home for dinner – "Don't lean your elbows on the table", "Sit still", "Don't talk and eat at the same time", "Always say please and thank you", "Don't shout".

Children cannot always be expected to fully understand why these rules exist, just that they need to follow them. But as they grow and develop they learn that the rules they once knew can be treated as mere guidelines, this is because they have learnt (hopefully) that the rules can now be replaced with something far superior – genuine heartfelt respect and courtesy for their fellow human beings.

Similarly, especially for adults, it is generally better if people behave morally not because the law dictates they should, but because they are personally motivated to do so.

## Inciting good behavior

The task of attempting to persuade us of the merits of a moral life has, throughout history, tended to belong to the province of religion. There are two main tactics which have been employed to do this.

One of these involves pointing out some advantages of good moral behavior which might be less than obvious. The notion that we can make our own lives more fulfilling and happy by extending our efforts for the betterment of others is not necessarily a religious one. Social studies have noted that people who, for instance, volunteer for charitable work tend to lead happier and healthier lives.

*A good example of this is the idea often encouraged by the Dalai Lama, the Tibetan Buddhist leader, of being "wisely selfish"...*

But down the ages some spiritual leaders, particularly in eastern religions, have picked up on this notion and promoted the idea that to be good is actually in your own self-interest. A good example of this is the idea often encouraged by the Dalai Lama, the Tibetan Buddhist leader, of being "wisely selfish" and being compassionate practitioners of the Buddhist path, as opposed to the "foolishly selfish", whose only interest is in themselves.

The other tactic involved is to motivate by referring to what is implied by the worldview of the religion in question (by worldview I mean the accepted notions of reality which are taught by a religion (examples include the existence of angels, the soul, miracles, life after death etc.) and tend to be accepted as a matter of faith).

*"Well it isn't easy to sleep," he told me in an unsettlingly matter of fact way, "when you know your wife is burning in hell!"*

I once had a work colleague whose demeanor was of a deeply unhappy and fractious man. I was having a lunchtime chat with this co-worker one day, and he seemed more haggard even than usual and seemed to be aware of this because he

than usual and seemed to be aware of this because he

147

made a point of explaining to me he was tired. I asked him if he had had a late night. "Yes," he said, "but I don't sleep well at the best of times." I asked if there was a reason for this. "Well it isn't easy to sleep," he told me in an unsettlingly matter of fact way, "when you know your wife is burning in hell!" Aghast and utterly taken aback I asked him how he could be so sure of such a thing. "I know," he insisted, "she could never have gone to heaven".

The source of this poor man's anguish was derived from a tragic mix of his catholic convictions, having been in love with someone who did not share his faith and had sadly died, and a little bit of unfortunate logic.

The belief in heaven and hell is quite a powerful example of a religious worldview which can be employed to try and motivate people to behave well (though in the instance quoted the effect can have worrying consequences), and is effective if the picture is completed by an all seeing all powerful god who, after we have died, will choose whether we should spend the rest of eternity in heaven or hell. This is an idea probably most associated with Judaism, Christianity and Islam.

Believing this might certainly be enough to encourage people to do the "right" thing in situations where they would otherwise have believed that nobody was watching, though we might wonder if it is worth the potential anguish.

Another worldview product is the concept of the principle of karma, which plays an important part in religions such as Hinduism, Buddhism, Jainism and Taoism. It is a complex idea and has subtle variations but broadly speaking it is a law of causation resulting from peoples actions.

This tends to mean that one's circumstances (good or bad) are the result of one's actions, whether recent or in a previous life (these religions also have the idea of reincarnation included in their worldview). In other words, simply put, if you do something bad now, you will suffer for it at some point; do something good now, and you will ultimately feel the benefit. This is regarded not so much the result of a judgmental god but rather a kind of independent physical law by which the universe works.

**The Yuwayol imperative**

Yuwayol, if taken seriously, also provides us with a worldview. This worldview does not imply any gods or other supernatural beliefs. Nor does it involve heaven, hell or karma. Nor does it really tell us much about how the universe (or totaliverse) works.

However it does assert that in order to exist, all things must be the same thing, and therefore we are all the same thing. My self and your self are the same.

> *...your welfare is in a literal sense the welfare of my own self.*

This clearly provides a moral motive in that I should be just as concerned by your welfare as I am my own, because your welfare is in a literal sense the welfare of my own self. If I was to punch you on the nose then the pain inflicted would be my pain just as much as if I'd punched myself on the nose; it's just that, as me, my self would not be so directly aware of this, but, as you, my self most certainly would be!

With Yuwayol the instruction quoted at the beginning of this chapter, to 'love your neighbor as yourself', takes on an emphasized twist. You should love you neighbor as yourself not just because she/he has equal value to yourself, *but also because his or her self is your own self.* From this perspective it can be seen that Yuwayol provides us with a new "selfish" interest in the wellbeing of others.

And since Yuwayol says that we are all aspects of the same thing, consequently our moral horizon has at least in theory expanded to include all living things, wherever they may be and whatever their relationship might be (or not be) to us.

### None of your business

So, having discovered a new incentive to take an interest in your fellow humans' welfare, perhaps you are wondering if this gives you a right to interfere with other people's lives, even if perhaps that interference is unwelcome. If their happiness is your happiness then that gives you every right to tell them how they live their lives, right?

Well no, I don't think so.

There is a marked difference between concerned assistance and interference. We are all uniquely responsible for ourselves and the way we live our lives. We are

> *...you, not I, should be the arbitrator of how you live your life.*

familiar with our own likes and dislikes, strengths and weaknesses, our goals and ambitions. Yuwayol changes none of this. As me, my self cannot understand the way my self, as you, needs to live your life better than you do. For this reason you, not I, should be the arbitrator of how you live your life.

That is not to say that we cannot help others when that help is sought or needed. With or without Yuwayol, we should endeavor to use good judgement when deciding when to lend a helping hand or when to step back, just as we always ought to have done.

**The same rules apply**

This same logic can in fact be applied to all moral issues so far as I can tell. So long as you regard morality to be about how you treat yourself and your fellow travelers in life, rather than as a set of rules, then Yuwayol should not change your views on what is, and is not, moral.

> *So long as you regard morality to be about yourself and your fellow travelers in life... then Yuwayol should not change your views on what is, and is not, moral.*

151

But having said that, depending on how much credence you give to Yuwayol, it may change your attitude to your own moral behavior.

To illustrate, let's imagine a couple of characters whom we will call Ann and Bob. Now let us say that Ann and Bob are both people who share the point of view that the state should not grant financial aid to people out of work. However they have very different reasons for thinking like this.

Here are their reasons:

*Ann has a lot of truck with the concept of 'tough love'. She feels that giving people aid for not doing any work provides a disincentive for the population as a whole to go out and find work, this in turn makes people lazy and therefore ultimately does more harm than good. She is sincere in this belief and would most likely change her mind if she could be convinced that giving benefits would make things better rather than worse.*

*Bob on the other hand, doesn't really go along with Ann's view point, but actually he couldn't care less about people who are out of work. Though he feels, in his heart of hearts, that financial aid would help people in a tough spot, this doesn't matter, because he realizes that giving out benefits means higher taxes for him and he quite simply doesn't want to pay higher taxes and that's the end of it.*

Now let us say that one day Ann and Bob both read about and are persuaded that Yuwayol is very likely to be correct, and that their self is something that they share with all people, including those who are out of work and wishing to receive financial aid.

In this situation Ann should not change her mind. Even though she is now aware that the out of work people's pain is in fact her own pain, and even though this might increase her distress, this says nothing which addresses her sincere belief that to give financial aid simply makes things worse for everyone overall.

On the other hand we may well expect Bob to change his view. He should realize that those seeking benefit have an equal right to his considerations than he himself does. Their suffering is his and, because he doesn't really go along with Ann's tough love ideas, he might well consider that alleviating this pain is worth the cost of an increase in his tax bill.

Now I don't want to get diverted by discussing who is right about benefit payments, Ann or Bob. The point that I'm making is that, depending on the degree you accept the validity of Yuwayol, your motivation for making sound moral judgements as honestly as you can should only increase according to how little concern you previously had for your fellow beings.

It is an important point – with Yuwayol what is considered right and wrong is just as much a matter of honest judgement and opinion as it ever was.

# Chapter 10

# Is Yuwayol a Religious Thing

*"Hope burns eternal in the human heart"*
***O.R. Melling***

*"Yes I am, I am also a Muslim, a Christian, a Buddhist, and a Jew."*
***Mahatma Gandhi (when asked if he was a Hindu)***

### Defining religion

A friend of mine was quite put out one day when I said something (I cannot remember what it was) which suggested to him that I did not think he was a Christian.

Not wishing to offend I was quick to apologize and explained that I had had no idea that he harbored any religious convictions. "I don't," he told me. This rather confused me so I enquired further. Did he for instance, I asked, believe that Jesus Christ had been the son of God and that he had died for our sins to be resurrected on the third day? "Of course not," he replied scornfully, adding that he didn't hold any truck with any of "that kind of nonsense".

I continued to probe and it turned out that he never read the bible, never went to church and seemed to have virtually no knowledge of the teaching of Jesus whatsoever. In the end I eventually realized that his definition of a 'Christian' could roughly be described as someone who was a pretty morally decent sort of person.

Now I have met many folk in my life whom I would happily describe as pretty morally decent, but a number of them would be mortified if I was to describe them as Christian.

It seems to me that such differing standpoints occur because religious sounding words such as 'Christian', 'devout', 'sacred', 'faith' etc. often mean different things to different people. This I think is particularly true of the word 'religion' itself. In the chapter, "*Moral Imperatives*", I touched upon the relationship between religion and morality and pointed out how Yuwayol offered a moral imperative in a not so dissimilar way to the way religion can.

> *...religious sounding words such as 'Christian', 'devout', 'sacred', 'faith' etc. often mean different things to different people.*

So does that make Yuwayol a religion? Well I'd say no (how's that for a quick answer!?), but I think it is worth explaining why whilst taking a look at the way different people see religion and its purpose.

**World view**

One thing that seems common to all religions is that they all provide stories and beliefs, which have been passed down from generation to generation, sometimes in written form, sometimes orally. These stories typically convey certain ideas about the world, such as its origins, and will often attempt to explain fundamental concepts of how things work.

*Most religions have their own creation stories which aim to explain how the universe came into being.*

Most religions have their own creation stories which aim to explain how the universe came into being. The story told in both Christianity and Judaism is found within the book of Genesis which tells the story of how God creates both the heaven and the earth in six 'days'. In Islam Muslims refer to the Qur'an which tells a similar story.

For many people the belief in a world view is what defines a religion. This is what caused my confusion when my friend told me that he considered himself to be a Christian, even though he did not hold with what many (myself included) would presume to be Christianity's core beliefs, that the man they call Christ (a certain Jesus who lived in Judea from around 5 B.C to 30 A.D) was the son of God and was resurrected (came back to life) after having being crucified to death. I think I would be right in saying that a very large proportion of those who refer to themselves as Christians would agree that if you don't believe this then really you aren't actually a Christian.

It is important here to understand what we mean by 'believe' as this is a word one might use more lightly when, for instance, discussing the weather – "I do believe it will rain this afternoon". In this context one is simply stating that it seems likely to rain, a best guess if you will, most likely based on the current local climatic conditions.

However, in a religious context the word 'believe' takes on a much more emphatic meaning. A religious belief is something that one has as a matter of conviction, something to feel passionate about, to defend and even, in some circumstances, something some may be prepared to die for (and in some highly disillusioned and tragic circumstances, kill for).

The justification for this type of belief is not based on any real evidence, at least not the sort of evidence you might expect to stand up in a criminal court. Rather it tends to be based on the notion of faith, which is often explained in terms such as "knowing within my heart".

For example, even today in the scientifically and technologically advanced western world, there are many who dismiss the scientifically backed theory of evolution by means of natural selection simply on the grounds that they feel it contradicts the Genesis story of creation.

Now when it comes to Yuwayol I really cannot say that I "know in my heart" that it is the truth. Which is one reason why Yuwayol is not a religion. Yuwayol is a philosophy which makes sense to me. It is my best guess. If I were forced to place a bet as to what my place in the world (or

> *...when it comes to Yuwayol I really cannot say that I "know in my heart" that it is the truth.*

totaliverse) is, I would put my money on the ideas I'm putting forward in this book.

Some people say that you have to believe in something. I'm not sure I agree. But if you do feel you need to believe in something, Yuwayol is the belief I'd recommend. I just ask that you don't look to go around dying or killing for it that's all!

**Playing by rules and traditions**

For some though, believing in a particular world view is not essentially what religion is all about. Whilst chatting one day with a lady I once knew, she told me that she tended to regard her membership of the Roman Catholic faith in the same sort of way as she did to a membership of a golf club. "In both cases," she told me, "If you want to be a member, you have to play by the rules."

She is one of many, I think, who feel that things like ceremony and lifestyle, the practices and the set of rules that one follows, are the things which define and are important in a religion, rather than the set of specific beliefs that you have.

The sort of practices I am talking about might include things such as going to a common place of worship on a regular basis, praying, meditating, circumcision, fasting, communion, sacrificial ceremonies, various wedding practices, wearing of headscarves, washing feet, avoiding particular foods at particular times, feasting at others especially when celebrating religious festivals, observing the Sabbath, baptism, and many more. Whilst rules might include anything from serious things such as desisting from violent behavior, to not eating 'unclean' foods or abstaining from drugs or alcohol, down to trivialities such as eating fish on Fridays.

As discussed in "*Moral Imperatives*", whilst Yuwayol does provide motive for moral behavior (assuming we define moral behavior as that which results from wanting the wellbeing of others as much as we want it for ourselves) it does not of itself throw up any particular rules or practices.

## Spirituality

A common thread which runs through the central message of almost all of the world's major religions is the notion of pursuing a spiritual path, as opposed to pursuing what might be described as more 'worldly' pursuits.

I think it is the strength of this commonality which Ghandi most likely had in mind when he said what I have quoted him saying at the start of this chapter, that as well as being a Hindu he also could claim to be "a Muslim, a Christian, a Buddhist and a Jew". Anyone (whether they are religious or not) who tends to see religion as a set of beliefs or rules and practices would probably find Ghandi's comments absurd, but there are many others who are happy to go along with his way of thinking. That is that most of the world's religions are, at some deep level, saying the same thing though in different ways.

> *... 'spiritual' is often regarded as being representative of a non-physical reality which is hidden from us...*

In this context the word 'spiritual' is often regarded as being representative of a non-physical reality which is hidden from us behind, as it were, the physical world that we see. So we are deceived, one might say, if we think that the way we perceive the world tells us everything about it. The claim being that the non-physical world is more real and eternal than the physical world, which is transient and therefore not to be overly relied upon.

For instance, our bodies can be regarded as belonging to the temporary physical world we are familiar with, and as such cannot last forever, which is why we die. On the other hand our inner spirit or soul, it is claimed, is (or has the potential to be), immortal.

So, the argument generally goes, one's focus in life should be to build up and strengthen one's spiritual side rather than aiming to indulge one's physical urges which will ultimately gain you very little.

A spiritual path is therefore one where practices such as prayer, meditation, love and compassion to others (which are deemed spiritual activities) takes preference over pleasure, greed and self-gratification (which are physical indulgences). Though it is normally, but not always, stressed that physical pleasure should not be abandoned altogether, rather that it should not take preference over more spiritual undertakings.

The idea of an actual spiritual realm, separate from the physical realm, is not really compatible with Yuwayol. Yuwayol says that every thing is the same thing. The totaliverse, in all its aspects, is a single entity so nothing can be separate.

Interestingly though, I think Yuwayol does go along with the idea that the way we perceive our world through our senses does mislead us in some ways as to the nature of the world we live in.

In particular, as conscious human beings it is far from obvious to us that each of our selves are in fact the same entity, that my self is the same as your self and everybody else's self. But Yuwayol does make that claim, and if that does inspire people to conduct their lives in any way differently (such as in realizing they have as much interest in other people's needs as they do their own) then I guess a parallel can be drawn with religious notions of spirituality.

But we need to be careful. 'Spiritual' implies a supernatural element in many peoples thinking. Yuwayol does not.

Actually I'm not at all sure that a belief in the supernatural is a necessary requirement for a spiritual lifestyle. Practicing meditation, for instance, which can help us to focus and assist us in keeping a calm and balanced view of our lives, does not really require a belief in another unseen world. Nor does living a caring, compassionate and meaningful life. Also, a beautiful landscape can be appreciated whatever one's beliefs.

The way I see it is that on one hand nature has bestowed upon us various needs and desires whilst on the other it has also built us in such a way that we will be happy and fulfilled in certain circumstances. However it can be a mistake to presume these two things are always correlated.

In other words I think it is wise to be mindful that what we want, and what makes us happy, are often two different things. You don't need to believe in a spiritual other-world to appreciate that.

**Is nothing sacred?**

Another feature of religion is the notion of certain things being considered sacred. Anything, be it an object, a dance, a story or a piece of music might be considered especially 'holy'.

This is because the *thing* in question is typically considered to have a close connection with some spiritual entity, such as a god.

In some cases this might elevate the thing in question in the minds of the faithful so that it effectively takes on a greater value than even the life of a fellow human. People can and do kill and die for the sake of these sacred things.

*In Yuwayol nothing is sacred. Every thing is the same thing.*

In this respect there is certainly nothing religious about Yuwayol as it has nothing to do with any of this. In Yuwayol nothing is sacred. Every thing is the same thing.

### The afterlife

Most of us fear death.

This is to be expected as it is in line with evolutionary theory that, in order to be motivated to survive a harsh and often hostile world long enough to breed and pass on their genes, our ancestors would have evolved to abhor the idea of their own non-existence. For humans though this fear of death has a particular poignancy as our intelligence has made us starkly aware of our own mortality.

It seems a cruel trick that nature has played upon us, that the thing we fear perhaps the most is the very thing we are fully aware to be inevitable. We will all die one day. Not just ourselves individually, but our loved ones also.

And when those who are nearest and dearest die, and as we struggle to come to terms with the stark truth that we will never interact with them ever again, that they are gone forever, that perhaps is when the fact of our mortality becomes the most apparent. There is here a melancholy sorrow which has always been a major contributor to the human condition.

*...most religions tells us that after death we can live on after all...*

It is as we come to terms with death that perhaps religion has its greatest hold upon many of us. Because most religions tells us that after death we can live on after all, that we are eternal.

In the Abrahamic religions (primarily Judaism, Christianity and Islam) the belief is that we (usually with certain conditions applied) have a future after death of an existence in heaven, a place of bliss and perfection. Other religions believe that, if not a heaven, then we do live on in some form of spirit world. Many eastern religions on the other hand claim that, after death, we reincarnate in a new body and that we have many lives, steadily progressing to an ultimate perfection.

Hope is a powerful thing. And for many, taking solace in the belief of a life after this one, for themselves and those they love, is both reassuring and helps them cope with a life that can feel sometimes cruel, pointless and futile.

But others take an opposing view. As there is no evidence of an afterlife we need to be courageous, they say, be dignified and face up to our mortality, only then can we hope be able to deal with our fear of passing away.

Perhaps the most significant conclusion of Yuwayol is that your self is not merely you, your self is in fact all living things (as well as all non-living things). So Yuwayol does have in common with most religions the notion that there is for us a life beyond our own.

I suppose this does lay Yuwayol open to accusations of wishful thinking. However, I'm not sure that what Yuwayol is suggesting is really what religious people tend to hope for.

*... Yuwayol regards our experience of time as a kind of illusory construct devised within the brain, the very notion of a 'life-after-death' is a kind of oxymoron ...*

As far as Yuwayol is concerned, when you die your memories will die with you. Any ideas or secrets you have not shared will be gone too, along with all your unique characteristics and the ways you have of seeing the world. There will be no joyous reunion of once lost loved ones to look forward to in some afterlife.

In fact, because Yuwayol regards our experience of time as a kind of illusory construct devised within the brain, the very notion of a 'life-after-death' is a kind of oxymoron – without the brain to create this construct there is no flow of time and therefore no 'after' at all when the brain has ceased to function.

However, the conclusion of Yuwayol is that your self, that significant 'you' we discussed in the chapter "*What am I (The Question)?*", being the entirety of the totaliverse, certainly does exist beyond the individual you who will one day die.

This is because you are merely one aspect of the entire totaliverse. Your self has lived, and continues to live, countless billions of lives, perhaps an infinite number of lives. I guess we might think of it as not dissimilar to reincarnation. Indeed, Yuwayol might be thought of as reincarnation gone extreme, but without any promise of nirvana to look forward to.

### In conclusion

So to the question "Is Yuwayol a religious thing?" I think the answer is – not really.

Yuwayol certainly proposes a world view – that all things are different aspects of the same thing, including you and me. However I would stress that no one is asking anyone to accept this world view as a matter of blind faith. One day perhaps science, rather than religion, might somehow unveil the truth of Yuwayol to us. Until then I would suggest that we can only look at what makes sense to each of us, keeping an open mind but reserving judgement.

In terms of rules and traditions Yuwayol has nothing to offer which is new. There are certainly no traditions associated with Yuwayol. This is probably because as I write I know of only one adherent to this very new philosophy, and that's me! Nor are there any rules, merely an incentive I think to do right by others, driven by a realization that they are, in fact, your self.

And whilst Yuwayol firmly rejects the idea of a spiritual realm separate from the world we inhabit, it does go along with the idea that the totaliverse is not what it seems. Nor does it invalidate practices which many associate with spiritual activity, such as striving for a meaningful life in preference to one beset with self-indulgence.

In a limited way Yuwayol does join with many religions in that it agrees there is a life for us beyond our own, many lives in fact. But those lives are forever separate from the one you are now experiencing. And Yuwayol differs from those other religions in that these other lives are not dependent on what you believe or what you hold to be sacred.

And personally I don't think that is a bad thing.

# Chapter 11

# And if Yuwayol is Wrong?

*"It is the mark of an educated mind to be able to entertain a thought without accepting it."*
**Aristotle**

*"Don't believe anything you read on the net. Except this. Well, including this, I suppose."*
**Douglas Adams**

### Life and soul

I can easily imagine that there are many who have read this book, and you may be amongst them, who will object to Yuwayol on the grounds that it ignores another long established idea relating to the nature of self. For many have concluded that we each have an individual 'soul' which can be distinguished from, and is ultimately separate from the physical world we see. This soul, the argument goes, is actually the real and essential self which, many believe, has the potential to survive our bodily death and live on in perpetuity.

You may consider this to be a better explanation, but it is one which I find difficult to accept for various reasons. One important reason is that it suggests a dualistic approach to nature.

Dualism in this context means accepting the view that the mind (soul) and matter are fundamental entities which are essentially separate.

My objection to this is that these two entities of mind and matter cannot be entirely separate from each other if one is to have any influence over the other. Our minds can clearly control our bodies, so there must be some sort of interface between the two, a bridge if you like between them.

If this is so then some physical manifestation of this interface should be available for scientific scrutiny, and so far no evidence for any such thing has shown itself.

Of course this does not have to mean that such a bridge does not exist but in any case such a proposition makes no sense to me. If the spiritual and the physical can interact with one another then by definition they surely cannot be separate.

*If the spiritual and the physical can interact with one another then by definition they surely cannot be separate.*

For me Yuwayol trumps the theory of a soul because Yuwayol does not require this dualistic approach. As I have stated on at least a few occasions in this book, any kind of notion of a separate supernatural realm is contradictory to Yuwayol because Yuwayol says that the totaliverse (which we have defined as being everything that exists and ever

172

has or will exist) is in fact a single entity, you and I and everyone else and everything else, including my teapot, are different aspects of that single entity.

However, an attractive inducement to believing that the soul is real is that it holds out the possibility that, through our soul, we can escape death and live forever. This is an attractive idea for many, especially as many religions promise that what is on offer is a life in eternal paradise. For anyone who has bought into this idea, any other worldviews are going to feel like quite a let-down.

But of course we cannot simply accept something as true simply because it sounds good. I think it would be jolly nice if I was to win the National Lottery this week but that doesn't mean that it is going to happen (even if I was to get around to buying a ticket!).

All this been said, I know it would be naïve of me to think that my arguments are so persuasive that everyone who believes they have a soul is going to drop that belief right now. The argument is endless. For such people such beliefs are as much a matter of faith as they are logic.

And hey! What do I know? Maybe they are right after all. I just doubt it that's all.

But for now I need to be reasonably succinct and hope you will forgive me if I proceed as though the argument is settled. For the remainder of this chapter I shall presume that there is no soul.

## Here and gone

I have tried throughout this book to make the case for Yuwayol. I have rested the case by means of the consideration of two historically difficult and hotly contested philosophical questions, 'What am I?' and 'Why does anything exist?'

I admit the idea does run contrary to what we generally consider to be our common sense notions of reality. Indeed part of my argument for Yuwayol includes the notion that the way we are hard-wired to perceive the world around us is misleading in regard to the nature of time and space.

I have also concluded, for reasons which I have given, that what we consider to be our essential self must be a real thing, and not some sort of abstract offshoot produced from our conscious mind.

## But what if that conclusion is wrong?

It takes roughly a second for the human brain to perform the many functions needed in order to perform what is probably its cleverest trick and generate each moment of consciousness. These moments might overlap somewhat, and they share common memories and feelings etc., but they are all special and unique, having been produced by a subtly changed brain on each occasion. They stand alone and are identifiable.

Past moments are mere memories; future moments are yet to happen.

Which means the you who is reading this sentence is not the you who started reading the chapter to which it belongs, nor are you the you who will (I hope) finish reading it.

*...the you who is reading this sentence is not the you who started reading the chapter to which it belongs...*

If the essential you that we spoke about in the chapter "*What am I (The Question)?*" really is some sort of conjured entity then you will be gone in the next second or so to be replaced by another entity which will inherit your memories and feelings. And she/he will be gone a second or so after that. And so on... each 'you' convinced by the same illusion of fluid like continuity.

If this sounds depressing then perhaps you should consider the supreme oddness of your existence in the first place. Viewing the world this way it is hard to understand how you come to be here at all and, well, "easy come easy go," as they say.

### Yuwayol-lite

This view of our self is completely the opposite of the one offered by Yuwayol. Whereas Yuwayol says that your self is the entire totaliverse, a very grand claim indeed, in this other view your self is merely a fleeting conscious moment of, well, you.

And yet it seems to me that as polarized as they are, these two views have certain things in common. So much so that I am inclined to apply the moniker 'Yuwayol-lite' to this fleeting notion of self. This is because by following through on either philosophy Yuwayol and Yuwayol-lite both deliver, in a peculiarly paradoxical way, a similar outlook.

As discussed in "Moral Imperatives", if Yuwayol is correct then I ought to have as much interest in others' wellbeing as I have for my own because my self is those others. Conversely if Yuwayol-lite is correct then I still ought to have as much interest in others' wellbeing as I have for my own but for a different reason, namely that my self is about to pass away so my future should be no more important to me than anyone else's.

*...if we all don't particularly exist, then can we not conclude that we all are the same, non-existent, thing?*

Curiously, a kind of universality still applies. I (and by 'I' here I mean my identity which my self is only a current fleeting moment of) and everyone else are common in that our selves don't in anyway meaningfully exist. After all, if we all don't really exist, then can we not conclude that we all are the same, non-existent, thing?

If you think this all feels a little nonsensical I am inclined to sympathize. For myself, Yuwayol-lite has an untidy feel about it and, if only for this reason, I favor Yuwayol proper.

**And finally**

I have spent a large, some might say unhealthy, proportion of my life wondering about what my relationship with the cosmos is. And it is with some bemusement that I have to conclude that for all my contemplation on the subject, I have arrived at no conclusive answers nor, I suspect, am I likely to do so. And I have come to regard with suspicion any who claim to have expertise in such matters.

Having said that, for me Yuwayol makes sense because, whilst it doesn't entirely answer the two most profound questions we can ask ('what are we?', 'why does anything exist?'), by linking them together it removes their paradoxical nature, and so it has become my best guess. In fact it seems to me to be the only reasonable guess I can make.

Undeniably, the conclusion of Yuwayol, though simple, is dramatic – through you the totaliverse becomes alive. But it is a bitter-sweet philosophy.

If Yuwayol is true then your self encompasses all things and all other creatures great and small, extending the scope of your being to a greater extent than any of us can ever imagine; at the same time each individual life, with all its precious dramas and beauty, that your self will experience are each finite and tiny, fundamentally lonely and sundered from one another.

I suggest you make the best of the life you lead now.

How you feel about this is a matter for you to decide. I have no desire to preach and would always advise you to be skeptical. To be skeptical is to be truly open minded, for it entails the questioning of all things, including oneself. (This is opposed to the cynic, who by wishing to be nobody's fool, has a nature dedicated to mere rejection.)

If, after having thought it all through, you have been persuaded that there could well be something to Yuwayol, or alternatively if you find yourself unconvinced, whichever, then I hope that you have at least found some interest in these pages, or even that they have provided you with some food for thought.

Thank you for reading.